A Century of Radiation and Medical Physics at New York's Memorial Hospital

A Century of Radiation and Medical Physics at New York's Memorial Hospital

Judith Groch, John Laughlin, Jean St. Germain, and Clifton Ling

with Lowell Anderson, Chandra Burman, Jenghwa Chang,
Chen-Shou Chui, Lawrence T. Dauer, Edward Epp,
Doracy P. Fontenla, John Humm, Andrew Jackson,
Gloria C. Li, Michael Lovelock, Gig Mageras,
Radhe Mohan, Lawrence N. Rothenberg,
Ellen Davis Yorke, and Marco Zaider

A Century of Radiation and Medical Physics at New York's Memorial Hospital
© 2019 by Judith Groch and C. Clifton Ling

Library of Congress Control Number: 2018966759
ISBN soft cover: 978-0-944838-08-2
ISBN eBook: 978-0-944838-41-9

Medical Physics Publishing
4555 Helgesen Drive
Madison, WI 53718
Phone: 800-442-5778
Fax: 608-224-5016
www.medicalphysics.org

Printed in the United States of America

Contents

Dedication

To all the members of the Medical Physics Department
during the first century of radiation and medical physics
in New York's Memorial Hospital

Preface

The discovery of x-rays by Wilhelm Röntgen and radium by Marie Curie marked the beginning of the contributions by physicists to the field of diagnostic and therapeutic radiology. Almost immediately after those discoveries, astute physicians started using x-rays and radium for the diagnosis and treatment of cancer. Physicists who worked side-by-side with physicians were pioneers in the field that became known as "medical physics." Over the last century, the specialties of radiology and radiation oncology have advanced significantly, due in part to the contributions of medical physicists.

With the appointments of James Ewing, Henry Janeway, and Gioacchino Failla to the Memorial Hospital for the Study of Cancer and Allied Disease (now known as "Memorial Sloan Kettering Cancer Center") in the early 1910s, the staff of Memorial Hospital became deeply involved in the development and application of radiation in the treatment of cancer. This short treatise provides a glimpse of the activities and contributions of the medical physics staff over the span of one century, approximately 1910–2010.

In the 1990s, John S. Laughlin, Ph.D., began assembling documents from the Memorial Hospital archive to write a historical account of the use of radiation and medical physics. His effort was only partially accomplished at the time of his untimely death in 2004.

Jean St. Germain, M.S. (1945–2017), who worked at Memorial Hospital for 50 years, together with C. Clifton Ling, Ph.D., continued that effort. To achieve a polished final product, they recruited a professional author, editor, and dear friend, Judith Groch Minowitz, as the writer and editor.

In addition to the efforts of John, Jean, Clif, and Judith, other medical physicists in the Department of Medical Physics contributed to the writing of many chapters, as noted in the authors list.

Finally, whereas this book represents the best of our collective efforts, there are likely unintended inaccuracies and omissions, for which we offer our apology.

Judith Groch
C. Clifton Ling

Before the Beginning

On Central Park West in New York City between 106th and 105th streets stands a luxury condominium: 455 Central Park West. Its massive round towers suggest a French château. Steps from the park, with all the features of a luxury dwelling and units selling for millions of dollars, the building lacks only the appropriate scenery to set it on the Loire in France.

But it was not always so.

The building has a storied history going back to 1884, when it was founded on 106th Street as the New York Cancer Hospital. Completed in 1887, the hospital was the first in the country devoted to the treatment of cancer patients. The building's big round towers were designed to prevent germs and dirt from accumulating in sharp corners, believed at the time to harbor disease. An air shaft, which ran vertically through the center of each tower, was considered the latest in ventilation technology.

Since the development of anesthesia in the mid-19th century, surgery had become the single treatment for cancer, thereby creating a major role for surgeons in the years to come. At that time, however, cancer treatment was mainly palliative, with patients coming to the hospital for morphine and to die.

Unfortunately, given cancer's deadly nature and the stigma of the disease, the hospital soon ran into financial trouble, so that at the turn of the century the hospital changed its name to the General Memorial Hospital. The hospital was used for general purposes, while gradually cancer patients were restricted to two wards partitioned off from the rest.

At that point, the unfortunate institution awaited, in a sense, the discovery of radium by the Curies in 1898, research and engineering advances, and the brilliant, dedicated pioneers—among them James Ewing and Gioacchino Failla—who would build the foundation of radiation physics.

James Ewing, Radium, and the Birth of Medical Physics

A major step in diagnosing disease came with the discovery of x-rays by Wilhelm Conrad Röentgen (1845–1923) in 1895. In 1896, Henri Antoine Becquerel (1852–1908) found that uranium salts without exposure to sunlight spontaneously emitted invisible rays, similar to x-rays. These rays could affect a photographic plate after passing through thin plates of metal. Yet having discovered radioactivity, Becquerel stopped there. For his early researches on radioactivity, Becquerel later shared the 1903 Nobel Prize with Marie and Pierre Curie.

Then in 1897 Marie Sklodowska-Curie (1867–1934) and Pierre Curie (1859–1906) began to study Becquerel's rays and took the next step. After investigating 13 elements—such as gold and copper, which did not give off rays—Marie Curie made a stunning discovery. A sample of pitchblende, the heavy black mineral compound known to be a source or uranium, appeared to give off rays similar to those Becquerel has observed from uranium but much stronger. Working through a ton of Austrian pitchblende, the Curies finally identified two radioactive elements, polonium named after Marie's homeland, and radium, a decay product of the uranium ore first discovered in the Austrian Sudetenland in 1789.

It took until 1902, however, before the Curies, working tirelessly, succeeded in extracting 100 mg of the new element from a ton of Bohemian uranium ore, enough to determine its atomic weight of 225.

With its official acceptance, radium moved beyond experimental studies, opening a new page in the treatment of cancer, until then a disease stamped with death.

After the Curies' discovery, the Austrian government embargoed the uranium–radium pitchblende ore. Meanwhile, Paris refineries, cut off from Austrian pitchblende ore (50% uranium), turned to low-grade vanadium ore, which French geologists had discovered in Utah and Colorado in 1899.

3

The pitchblende ore mined in Colorado was shipped to France for refinement so that American physicians were forced to buy radium from European refineries. Thus there was extreme pressure in the United States to find a way to break the European monopoly and preserve American radioactive sources.

James Ewing

At this point we pause for a glimpse of James Ewing (1866–1943), a physician and founder of what would eventually be known as medical physics and of the Memorial Hospital we know today.

Born in Pittsburgh in 1866, Ewing was a brilliant and tireless worker, a scholar, and among the few to recognize at an early stage the potential of radiotherapy for cancer, which he called "the first rational treatment of cancer ever devised."

Ewing received his medical degree in 1891 from the College of Physicians and Surgeons at Columbia University in New York. After six months of surgical service at Western Pennsylvania Hospital in Pittsburgh, Ewing returned to New York for an internship at Roosevelt Hospital and Sloane Maternity (later named Sloane Hospital for Women), affiliated with Columbia University. In 1893, Ewing accepted a position as an instructor in histology at Columbia University and, following study in Germany, he was promoted to assistant in clinical pathology at Columbia.

After serving as a contract surgeon for the U.S. Army during the Spanish American War, Ewing, age 33, was appointed the first professor of pathology at Cornell University.

In 1902, the C. P. Huntington Fund for Cancer Research was established at the Alfred L. Loomis Laboratory at Bellevue Hospital under Ewing's supervision. In 1910, the Huntington fund sent Ewing to an International Congress of Cancer Research in Paris. With his growing interest in the field of cancer, Ewing became convinced that the current belief in the parasitic hypothesis was not valid for human cancer. To further cancer clinical research, Ewing approached New York Hospital proposing the appointment of a commission for clinical research, but the plan had problems and was withdrawn.

A New and Improved Hospital

That attempt, however, brought him into contact with James Douglas (1837–1918), a mining engineer and the wealthy president of the Phelps Dodge Company. Douglas's daughter suffered from recurrent breast cancer after a mastectomy. Radium treatments at Guy's Hospital in London with expensive radium bought from Paris were disappointing, and the young woman died in 1910. Douglas believed in Ewing's plan and asked Ewing to draw up a statement of resources. Excluding the Huntington fund, the total amount came to $2,400,000.

After the failed attempt at New York Hospital, Ewing and Douglas made an offer to the Board of Managers of the General Memorial Hospital to return the institution to the exclusive care of cancer patients and establish an affiliation with Cornell, an enduring relationship as it turned out.

Until then, the hospital was used for general purposes, while cancer patients were restricted to two wards partitioned off from the rest. The ward partitions of the hospital were eventually removed and the word "General" was dropped from the hospital's name. In 1912, the institution became known as the Memorial Hospital for the Study of Cancer and Allied Disease. At this point, Ewing, serving as pathologist, became acting head of the newly named institution, assuming full responsibility for the hospital's policies.

In a fortunate event, Ewing and Douglas, who had been seeking to increase the radium supplies at Memorial, met Dr. Howard Kelly (1858–1943), a distinguished and wealthy American professor of gynecology at Johns Hopkins University, who was pressing Congress for government support in radium mining.

In 1912 and 1913, German gynecologists had reported amazing cures of uterine cancer using radium, radon, or mesothorium in doses of a gram or more. Kelly had started using topical radium for brachytherapy in 1908, and by 1913 he wanted to use large doses of radium externally as the German gynecologists had pioneered.

The Triumvirate and Their Gifts

Driven by his inability to get sufficient radium for large-dose treatment, Kelly and now Douglas, eager to finance the cancer battle, joined Charles L. Parsons (1867–1954), Chief of the U.S. Bureau of Mines.

Parsons, representing the U.S government, bought property in Colorado and took over the extraction of carnotite ore, which was not as rich as pitchblende, but more abundant. The ore was delivered to a private refinery built and run by Kelly and Douglas, who had each put up $75,000 to build the facility. The alliance of Kelly, Douglas, and Parsons was incorporated as the National Radium Institute, which had a contract with the federal government.

The affiliation with Cornell included a legacy of one third of a million dollars ($348,000) and several grams of radium contributed by Douglas that was then worth $180,000 per gram. By late 1916, the venture had produced

8.5 grams of radium at a cost of $39,000 a gram to be divided among the partners. Each partner received 3.75 grams of radium.

Kelly's share went to Johns Hopkins, while Parsons' share went to Harvard and other institutions around the country. Douglas donated his share to Memorial Hospital, starting in 1914. By 1917, the accumulated amount of radium given to Memorial came to 3.75 grams. The relatively large amount donated permitted 2 grams to be put into solution and into an extraction plant to be used for therapeutic purposes.

The Douglas Trusts

In 1914 and 1917, two trusts described Douglas's conditions for his donations to Memorial Hospital. Both trusts placed the use of radium and the medical staff at Memorial Hospital under Cornell Medical School's control—a relationship that endured until the 1960s.

In the *Trust of May 20, 1914*, Douglas gave $348,000 to the "General Memorial Hospital for the Treatment of Cancer and Allied Diseases." The hospital was to devote itself exclusively to the treatment of cancer and allied diseases, with such diseases being designated by the Dean of Cornell Medical School. Furthermore, the physicians and surgeons on Memorial's Board of Managers were to be approved first by the Council of Cornell Medical School. This, for example, meant that the chair of radiology was subject to approval by Cornell.

In case of emergency, should the hospital wished to treat patients with non-cancer conditions, the hospital might do so, provided the Cornell Dean granted permission. Finally, should this trust fail or find it impossible to perform, the body of the trust was to be given to the trustees of Cornell University.

In the similar *Trust of January 29, 1917*, Douglas donated 2681.92 mg of radium with an estimated market value of $268,192 to the Memorial Hospital for the Treatment of Cancer and Allied Diseases. Of this amount, 307.20 mg were placed in solution in the radon plant.

The trust agreement stated that should Memorial have any radium to spare, it might be used by "friendly" U.S. hospitals, physicians, or scientists, given the written permission by the Dean of Cornell Medical School.

Basically, as in the earlier trust, the 1917 Trust left major decisions about the hospital's function, its leadership, and the use of radium to Cornell.

Furthermore, the document stated that if the trusts were to fail or should performance become impossible, then Douglas would give the radium to the trustees of Cornell University to devote to the study of cancer and allied diseases, and for general physical research. In addition, should any of the several limitations listed occur, the Memorial trustees and their successors were to turn the radium and the income from its use over to the trustees of Cornell University.

Again, Cornell oversight was to last for many years.

More Radium

In June 1917, the Department of Mines lent 689.6 mg of radium to Memorial Hospital, and in 1919, 194.5 mg of radium were added to the Douglas Trust amount, increasing the total to 2876.42 mg. In addition, Douglas announced that he would fund a complete rebuilding of the x-ray department.

In the 1920s, a $250,000 E. S Harkness donation for additional radium was used in the 4 gram radium pack.

In 1933, the Rockefeller Institute for Medical Research made a final donation of 803.20 mg. Thus a total of 3679.62 mg were donated between 1917 and 1933. Additional donations brought the total to 4614.4 mg.

William Duane Builds Memorial Radium Plant

William Duane (1872–1934), a former assistant to Mme. Curie, a professor of biophysics at Harvard, and a consultant biophysicist, had built a radium emanation, extraction, and purification plant to be used in his laboratory at Harvard. Now Duane agreed to install a model of his plant at Memorial.

Addressing the simple facts about radium and radon, Duane explained that in one gram of radium, about 37 billion times every second a high-energy alpha particle is emitted by an atom, which transmutes in a few days into a new element called radon. Radon, in turn, decays in a series of seven more daughters to other daughters, and eventually 23 years later to stable nonradioactive lead. Each disintegration, Duane explained, is accompanied by the emission of alpha rays (which cannot penetrate the skin) or beta rays (which penetrate only millimeters of tissue) and gamma rays, which are very penetrating.

In 1914 an International Committee in Brussels defined a "curie" as the amount of gamma emission from one gram of radium element sealed in a specific type of glass needle. Duane had substituted glass "seeds" for the steel needles. The glass radon containers (2 to 50 millicuries to a seed) were applied at various distances from the skin for weeks or months through various filtrations with astonishing results.

Henry Harrington Janeway

The Duane plant installed at Memorial was first used by Henry Harrington Janeway (1878–1921), a surgeon from New Brunswick, New Jersey, who was also skilled in developing laboratory procedures.

In building up the hospital, one of Ewing's tasks had been to recruit collaborators with a principal interest in malignant tumors. Janeway had published an article on skin cancer in a German journal. Janeway's work so impressed Ewing that in 1912 Ewing appointed Janeway as both chief of cancer surgery and of the new discipline of cancer radiotherapy at the still-named General Memorial Hospital.

In that position, Janeway and Douglas Quick (1891–1966), a young Canadian surgeon named attending roentgenologist, were charged with developing the techniques and equipment needed for interstitial and surface brachytherapy.

As far back as 1902, two low-voltage machines had been installed at the earlier Memorial hospital. Now in 1915, 464 patients were treated with x-ray sources from two new low-voltage x-ray machines installed in the operating room to treat residual or nonresectable tumors, an early form of intraoperative radiotherapy.

In the same year, a radium department was established at Memorial Hospital, treating 433 patients with Janeway as its director. Lewis G. Cole, a professor of radiology at Cornell, was the consulting roentgenologist.

Thus, Ewing, Janeway, Memorial Hospital surgeons, and the heads of the various subspecialties of gynecology, urology, and pharyngology became the curators of radium therapy in the United States.

During the years 1913 to 1916, Kelly and Janeway began using radium and radon to treat deep-seated cancers that required placing sources several centimeters away from the skin surface, so-called tele-curie therapy. Up to 4,000 millicuries of radium were housed in a lead-shielded cylinder suspended from the ceiling and moved by a pulley into position.

For this new technique, the surgeons needed several grams of radium, costing $180,000 per gram. Fortunately, the American surgeons were the beneficiaries of the National Radium Institute, founded by the original donors. With this source of radium and with Europe preoccupied in World War I, Kelly and Janeway developed megavoltage tele-curie therapy using the 1.2 mV gamma rays of "mass radium."

A main concern for Janeway was that the large amount of radium given to the hospital required housing, a means of protection, and methods of handling for clinical use. To manage this responsibility, he needed technical help.

In 1915, Janeway offered a part-time position to Gioacchino Failla (1891–1961), a young Italian-born electrical engineer skilled in mathematics, experimental physics, and with a knowledge of radioactivity. Failla plunged into the subject, and within a short time he learned to operate the plant and became knowledgeable about radioactivity and its medical uses. Soon everyone turned to him for information.

Now 65 years old, Ewing continued his efforts for Memorial Hospital. In 1932, having retired from Cornell Medical College where he still taught pathology, Ewing was made full-time director of Memorial Hospital, acquiring the title for a position he had held for many years and for the hospital he had helped build.

Busy as he was, Ewing's many diverse efforts extended beyond the two institutions he served. For example, in 1913 Ewing became one of the founders of the American Society for the Control of Cancer, the precursor of the American Cancer Society. Through his interest in research, he was also one of the organizers of the American Association for Cancer Research and served many terms as officer and council member. Furthermore, as early as 1918 Ewing published the first textbook on cancer pathology, "Neoplastic Diseases," which established a systematic basis for diagnosing cancer.

In 1920 Ewing had reported his observations of a malignant bone tumor that tended to occur in certain bones and frequently in teenagers, since then known as "Ewing's tumor." He reported that with radium treatment, the tumor receded and the shaft became well-defined with little deformity. Ewing's dislike of surgery seems to have stemmed from a childhood experience when a famous surgeon suggested amputation for the eight-year-old boy's right leg crippled with osteomyelitis. Fortunately, Ewing kept his leg, but the experience left him with a distaste for surgery.

Ewing retired as director in 1940 and was replaced by Dr. Cornelius P. Rhoads (1870–1959). In a 20-year period from 1913 to 1932, James Ewing built Memorial into the most influential cancer hospital in the world. He promoted his idea of cancer specialists with multidisciplinary training, insisting that surgeons and internists be trained to use radium, radon, and x-rays, an early idea that could not survive sophisticated advances in cancer therapy.

James Ewing died in New York City in 1943 of bladder cancer.

Gioacchino Failla, Inventor of Radiation Biophysics and Radiobiology

Much of the credit for the growth of radiation science at Memorial Hospital should go to the dedication and inventiveness of Gioacchino Failla (1891–1961), who over his long career converted physics into an independent service. In a short time, he went far beyond his earliest task of managing the radon plant and extracting radon for the clinical staff.

Born in 1891 in a small town in Sicily, Gioacchino Failla was raised by his physician grandfather after the death of his father when Gino was only three years old. Failla's mother, who had left to make a life in New York, brought her son to the city in 1906. Brilliant, inventive, and indefatigable, Failla earned an engineering degree in 1915, at which point Henry Harrington Janeway, who had been given the task of operating the radon plant, needed technical help and offered a part-time job to young Failla.

As a physicist, Failla's main responsibility was assisting with the supervision of the radon plant and developing new methods for using radon in cancer therapy. Gifted, with engineering training in his background, and totally dedicated, Failla tackled the subject, and in a short time he became an authority on the medical uses of radioactivity.

In 1917, Failla, Janeway, and Benjamin Stockwell Barringer (1877–1953), a Memorial Hospital urologist, coauthored a report on the early efforts at the hospital. Included was a long discussion of the physics of radioactivity by Failla and experience with radiation of patients with cancer of the skin, oral cavity, bladder, and prostate by the other coauthors.

In addition to managing the radon plant, Failla took an active part in the various trials with radioactive sources. For this purpose, he developed a machine shop in the basement of the hospital where he designed and built various accessories. Among his many gadgets was a bell-shaped lead container to hold the radon, which the hospital's gynecologists wished to bring into direct contact with the patient's cervix.

With the advent of the First World War, bilingual Failla was a natural choice in 1918 to become an assistant to the Scientific Attaché in the United States Embassy in Rome. After leaving service in 1919, he returned home via Paris, where he visited the Radium Institute and met Mme. Curie. She encouraged him to return and qualify for a doctorate from the University of Paris.

Edith Quimby Arrives

At Janeway's insistence, the hospital created a department of physics with Failla as its director. Then in 1919, Edith Hinckley Quimby (1891–1982) was added as his assistant, a fruitful collaboration that lasted for 40 years. An ideal match, they were responsible for the logistics of radium and radon. In 1922, when 140 kV roentgen therapy became available, Failla and Quimby oversaw its use. From the outset, Failla established a pattern of collaboration with his associates, as well as with medical colleagues, chemists, and biologists.

Quimby, who had a degree in physics from Whitman College in Walla Walla, Washington, had a talent for dealing with radiation problems, which earned her the respect of radiologists and physicists around the world. In the early years, radium-containing needles were applied to tumors with no certainty that the right exposures would be received. Quimby was the first to determine the distribution of radiation doses in tissue, using various arrangements of the radium needles. The techniques she described in 1932 for choosing the most effective grouping of the needles were widely adopted in the United States and served as the forerunner of Parker and Paterson's Manchester system.

During the same period, she quantified the different doses from beta and gamma radiation that produced the same biological effect, such as skin erythema, thus pioneering the concept of the *relative biological effectiveness* of radiation.

In later years, in addition to the studies of filtration and depth doses that Quimby did with Failla, she also undertook pioneering studies of time-dose relationships. In research for the Manhattan Project during World War II, Quimby produced recommendations for safeguarding individuals handling radioactive sodium and other newly available artificial radioisotopes, as well as for the safe handling of radioactive wastes and the clean-up of radiation spills.

Quimby's research also included the use of radioactive isotopes for treating thyroid disease, for circulation studies, and for diagnoses of brain tumors. These trials with radioactive sodium and iodine established Quimby as one of the pioneers of nuclear medicine.

Eventually as an educator, Quimby taught radiation physics and the clinical use of radioisotopes to new generations of physicians, physicists, radiology residents, and even industrialists.

Transition Years: 1918–1922

With the end of the war in 1918, Memorial's patient services resumed their annual increase. For example, both costs and receipts from patients more than doubled from 1917 to 1920. In addition, the hospital now had almost seven grams of radium, some of it owned by the Bureau of Mines and lent to the hospital.

In 1919, the hospital got a new chief of roentgenology, Ralph E. Herendeen, who stayed in that role for a quarter of a century. Lewis G. Cole remained as a consultant, adding Harry Imboden in 1921.

Henry Janeway—who had suffered for many years from an adamantinoma of the mandible and which he treated with radium—died at age 48 of either the bone tumor or of pulmonary tuberculosis in 1921. Janeway was among the first in the United States to advocate radium treatments as the treatment of choice for cancer of the cervix. He was also instrumental in developing equipment and applicators for radium, radon, and x-ray treatments. In his honor, the American Radium Society established an annual lecture in his memory.

Janeway was succeeded as radium physician-in-chief by Douglas Quick, his former assistant.

In 1921, Mme. Curie, a workaholic, took time to visit the United States, where she toured various cities and visited President Harding at the White House. At Memorial Hospital, Failla, who had known her from Paris, showed her the vault where the hospital kept its four grams of radium, as well as the hospital's procedures for using radium and radon seeds. Although Marie Curie had discovered radium by the arduous task of refining pitchblende, she did not have any radium and could not afford to buy any. A main purpose of her visit was to receive a gift of one gram of radium from a consortium of American women's groups, which had raised money to purchase it. She later traveled to Boston for a visit with Duane and his family.

Among the many papers Failla and his associates published was a 1921 paper on dosage in radium therapy. Failla emphasized the difference between the amount of energy available and that which was actually absorbed into the irradiated tissue. For an appraisal of results, Failla said only the dose absorbed by the tissue was important.

From there, Failla made an elaborate argument in favor of measuring the dose absorbed in calories. He calculated the microcalories absorbed in one cubic centimeter of tissue for various therapeutic procedures. The number averaged about four microcalories (on the order of 1700 rads in present-day data).

A 200 kV Machine, Radium, and Clinical Practice

In 1922, a new high-voltage x-ray machine became available with a peak beam energy of 200,000 volts. The machine was substantially stronger than previous x-rays sources, but still less potent than the gamma radiation from radiation emanation sources. In a 1922 report, Burton J. Lee, an attending surgeon on the breast cancer service, acknowledged the use of both radium and x-rays in breast treatments. The heads of several medical sections noted that the use of x-rays and radium grew both in volume and sophistication. Young physicians coming to Memorial were encouraged to learn how to use x-rays and radium to treat patients, with much of the instruction coming from the physics group, and particularly from Edith Quimby.

At the same time, Failla and Quimby started extensive dosimetry studies. For this purpose, they built an ionization chamber of bakelite and connected to a gold-leaf electroscope by means of a rubber insulating cable. Using a water phantom, they made elaborate studies establishing the depth doses in percentages of skin dose, using various types of filters and variable distances, followed by charts and tables for use in roentgentherapy.

Gradually Failla and Quimby extended their work into biophysics. They studied erythema of the skin in laboratory animals and patients, with the idea that erythema could be used as a biologic indicator with some degree of accuracy. For this work, they were awarded the Lenard Prize by the American Roentgen Ray Society, an award that honored Charles Lester Lenard, a distinguished radiology pioneer who had died in 1913 of radiation-related toxicity. Over time this award joined a page-long list of awards and honors in Failla's working history.

Gold Seeds

In 1923, Failla took a leave of absence to go to Paris and complete work for a doctorate under Mme. Curie. A year later he returned to the United States convinced that radiological research must include radiobiology. Adding biologists to his staff, he insisted that physicists associated with him also acquire some biological knowledge.

At this time, William Duane, who had installed the radium extraction and purification plant at Memorial, introduced the idea of fragmenting the capillary tubes into small sections. Failla built an apparatus that made possible uniform segmentation and calibration in a few minutes. Douglas Quick, a Memorial Hospital surgeon, and urologist Barringer used the bare "seeds" in a variety of tumors.

Although the radon seeds became very popular, a major problem was their lack of filtration and uneven irradiation of the affected tissues. In comparative tests, Quimby found that gold was an ideal filter, whereupon Failla developed the technological procedures to collect radon in capillary gold tubes that were adequately segmented and calibrated. The gold seeds served as a filter, permitting the use of gamma rays without the intense beta rays that usually caused inflammatory reactions.

In 1925, Failla went to London to participate in the First International Congress of Radiology. There an agreement was reached to appoint an International Commission on Units and Measures.

At this point, practitioners of radiotherapy who wished to relate their observed results to the amount of radiation administered thought that a biological unit might be the answer. On the basis of his own experimental observations, Failla dissected the problem and wrote that biological effects depended greatly on the wavelength of the radiations, that the various tissues reacted differently to the same amount of radiation, and that the time of delivery could not be excluded from the appraisal of effects.

Moreover, harking back to his earlier statement, Failla pointed out that there was a difference between the amounts delivered and the amounts absorbed. He concluded that air ionization, within a definable range of variables, was a more logical approach to establishing a unit of roentgen ray exposure.

In a 1926 report, Herendeen noted that the reason x-ray services had not increased over the previous three years was that the department was operating at its capacity and that in addition, the increase in long-exposure therapy had prevented the expansion of work.

Fortunately, this was soon remedied.

More Space

By the end of 1925, the hospital added a new building at 105th St. on the south end of the 106th St. property. Among its many improvements and new machines, the building provided space for two more high-voltage x-ray machines, several other machines, and overall, eight treatment tubes, plus a fluoroscopy room. Space was also provided for waiting and examination rooms, and on the 2nd floor more private rooms were installed to help carry the expense of the new building.

Development of Radium Teletherapy and Other Progress

The perfected techniques of brachytherapy had not solved the problem of treating large, deep-seated tumors. In addition to the high cost and lack of sufficient radium for tele-curie therapy, managing a relatively large source and a necessary short distance remained a problem.

By the middle 1920s, however, Memorial benefited from donations that enhanced its work, especially its growing research efforts. E. S. Harkness, the philanthropist son of a Rockefeller partner, donated $250,000 for additional radium, an amount to be used as a single treatment source, the so-called 4 gram radium "pack."

Failla built a unit to contain the pack. It had a "turn off" mechanism for protection of personnel, and the sources could be used at 8 cm or 10 cm from the portal. The heavy container was hoisted and mechanically transferred from one room to an adjacent one to allow for fast reuse from 7 a.m. to 11 p.m. The pack was used as an external gamma radiation source to penetrate tumor sites, much as high-voltage x-rays were used. The hospital's total radiation supply now came to eight grams.

Beginning his family's involvement with Memorial, John D. Rockefeller, Jr., donated the cost of new biological laboratories. In addition, in a significant new approach to cancer therapy, a fund was established in 1927 to support the investigation of chemical agents that might have anti-cancer effects.

By now, Memorial was receiving referrals from other physicians and direct appeals from potential patients. Interestingly, about 60% of the self-referred individuals were diagnosed as negative for cancer.

In July 1928, the roentgen as a unit of x-ray exposure was proposed by the International Commission on Units and Measures and approved by the Second International Congress of Radiology held in Stockholm. Failla and Quimby were both on working committees. Furthermore, the group concluded it was even more important for radiation users to have the beginning of standards for allowable occupational exposure.

In 1929, despite the stock market crash and the painful advance of the great depression, Memorial fared better than many other institutions. Wealthy board members—the Astors, Douglases, and Rockefellers, for example—rallied to keep Memorial growing.

The Energy Race—the Development of Higher-kVp X-ray Units

Experienced radiotherapists, going to extremes of target-skin distance and thickness of filters, had exhausted the therapeutic potential of 200 kVp radiation. Now a call came for higher-voltage units that would increase both penetration and the margin of safety between the effect on tumors and normal structures.

In 1932, William David Coolidge (1873–1975), who earlier had contributed the hot cathode tube, designed a 700 kVp unit that was installed at Memorial. Using a water phantom, Failla and his collaborators made careful ionization measurements for the 200 kVp and 700 kVp units, and the gamma rays for the large radium pack. Using wheat seedlings, drosophila eggs, and mouse tails as tests, Failla undertook a painstaking comparison of the effects from the two units and of gamma radiation from the 4 gm radium pack. At the same time, Quimby worked out tables of exposure time, source distances, and biological responses for the new machine.

These served as points of reference for comparison with several chemical and biological radiation effects, including clinical reactions. Fourteen collaborators from the laboratory and clinical staff of the hospital participated in the study. Precise numerical results could not be obtained, but the findings were essentially in agreement with results of similar studies carried on in later years with more refined methods.

Failla presented this study at the Fourth International Congress of Radiology in Zurich in 1934. Supporting the view of clinicians, Failla concluded that the quality of radiation played an important role in radiotherapy.

Higher-Energy and Lower-Energy Sources

The race for high-energy radiation sources continued. Now Memorial was selected by General Electric to test therapy with a new 900 kV machine. Failla worked with Coolidge and other members of the physics group to design a facility to house the new generator while developing new protocols for the powerful machine. Again, Quimby worked on tables of exposure time, source distances, and biological response to treatment. It was 1931 before the first patient was treated with the 900 kVp machine, although Failla chose to operate it at 700 kVp to 800 kVp.

In a 1932 report, Failla wrote that when operated at 700 kVp, the radiation was equivalent to the gamma ray emission of about 2000 grams of radium, as judged by the ability of the radiation to produce erythema of the human skin. Although far more penetrating than the 200 kVp machines, the new x-ray machine still fell short of the gamma rays produced by radium products.

The experience of the Memorial staff set national standards in many respects. The 900 kVp generator was normally used at 750 kVp and made a strong argument for the superiority of radiation in that range for many cancers. Herendeen and his colleagues handled 25,659 treatments during 1933. Part of this increase was the result of new treatment fractionation patterns, requiring more sessions to administer a selected dose. In addition, the volume of radium emanation uses for many types of cancer doubled over two years, while the four-gram pack continued in use.

A second approach was based on a concept by Arthur Heublein (1879–1932) who believed that certain patients (especially those with bone and lymphatic cancers) would benefit from two to three weeks of whole-body radiation at a tolerable level. The unit was used to treat advanced Hodgkin's disease, lymphomas, and other advanced cancers with good palliative results. The test required building a shielded set of rooms with a 185 kVp machine (often known as a Heublein unit) that could run almost indefinitely with automated controls and an automatic shut-off at the door.

Development of Radiation Protection Standards

One of Failla's major professional preoccupations was radiation protection, particularly for the physicians and technologists who handled radium. Over long periods of time, he made careful observations of the changes in their skin and nails, and he convinced himself that inexperienced and careless individuals exposed themselves the most.

In addition, a considerable number of the early workers in this field—including Mme. Curie, Eugene Wilson Caldwell, a physician and electrical engineer, Lenard, and possibly Janeway—were radiation's victims.

Although radiologists had previously used x-ray film to estimate radiation exposure, in 1923 Failla and Quimby were the first to start a full-scale film-badge program using dental film with a filter to distinguish exposure to gamma or beta radiation. Failla made the statement, commonly ignored, that in the concept of dose tolerance, time of exposure should not be excluded and that tolerance intensity should also be considered.

From a list of scientific articles published by the Memorial staff in 1925, some 49 dealt with radiation applications. Interestingly, an early harbinger of a new approach to treating cancer was the establishment in 1927 of a fund to support investigation of chemical agents that might have cancerocidal effects.

In 1928 an international gathering of radiologists and scientists met at the second International Congress of Radiology to develop an accepted set of definitions for the various radiation phenomena. It was even more important to have the beginning of standards for allowable occupational exposure.

Growing Demand for Treatment Results in New Problems

In 1922, James Duffy (1892–1942), a graduate of Harvard Medical School, headed the radiation therapy section. He was responsible for primary cancer treatment with the high-voltage machines and with the radium pack. At this point, Maurice Lenz joined the radiology staff, and biochemist Paul S. Henshaw joined the physics research staff. Failla and Quimby were regular examiners in physics for the new American Board of Radiology.

In 1934, as Memorial started its second half century of operation, the hospital had grown to meet new demands despite the economic effects of the depression. In 1935, the roentgenology department included two radiologists devoted to diagnostic studies, while two others plus two fellows and five technologists provided 26,484 radiation treatments. The physics department now grew to 10 persons, moved into new space, and added more technologists.

Despite the growth of radiology and physics, Memorial remained a hospital dominated by surgeons. Although diagnostic procedures were referred to Dr. Herendeen, the attending surgeon integrated radiation treatment with his surgery and maintained control of his patients.

Dr. Duffy was responsible for the physical performance of therapy treatments, but the referring surgeon determined the manner and amount of treatment and managed the patient during the treatment period.

The enormous demand for treatment prompted Memorial physicians to back away from extended treatment plans. From the earliest uses of radiation, it was clear that enough radiation could destroy any cancer. It was never possible, however, to identify the exact size, shape, and location of the cancer, nor whether the cancer had spread. For deep-seated tumors, external radiation had to pass through normal overlying tissues with enough energy to destroy the cancer below. Thus to enhance the ratio of radiation reaching the cancer versus healthy tissue, treatment plans included settings that reached the cancer from different directions, and even went further with the development of treatment sources that could rotate round a tumor axis.

One argument for protracted treatment was the growing understanding that cancer cells were especially sensitive to radiation at the time of mitosis. Thus the more daily fractions of radiation delivered to a cancer, the more likely that a maximum number of cancer cells would be destroyed.

The basic problem, however, was that by 1935 Memorial doctors were torn between the wisdom of more fractionation and the practical need to treat growing numbers of patients quickly. Because of the heavy load, shorter periods were used with doses seldom falling below 200 roentgen units. The results seemed to be as good as and sometimes even better than prolonged periods of treatment.

All told, however, the general belief was to use radiation as an adjunct to the surgical removal of a tumor, thus destroying cancer cells remaining in the tumor margins.

Occasionally, a case was made for radiation as a primary treatment. A senior Memorial surgeon, Frank E. Adair, argued for the value of treating even operable breast cancer primarily with radiation. Few of his colleagues agreed.

By the 1930s, the hospital's reputation had spread internationally, and its physicians had become leaders in their disciplines. In addition to Failla and Quimby, Lewis G. Cole and Harry M. Imboden remained consultants for almost 30 years. Ralph Herendeen and Duffy remained the on-site radiologists, responsible for both x-ray treatment and diagnostic procedures.

All U.S. radiologists during these years were certified and trained in both branches of radiology, with most concentrating on diagnosis. At Memorial, however, the demands of a cancer hospital pushed the radiologists toward therapy, even though other Memorial physicians thought themselves adequately trained to handle radiation treatment. Meanwhile, the Memorial troops soldiered on despite space and other limitations.

Fortunately, the limitations were about to end.

The Move to a New Site for Memorial Hospital

John D. Rockefeller, Jr., a member of Memorial's board, came to the rescue. Rockefeller agreed to donate a new hospital site in the Manhattan block between 67th and 68th streets and between First and York avenues. Others pledged money for constructing and outfitting the 12-story building. The site (400 East 68 St.) was across the corner of York Avenue from Cornell Medical School and New York Hospital, which had become Memorial's principal teaching affiliate. Memorial had been affiliated with Cornell for more than two decades, and new synergies were expected to arise from the proximity of the two hospitals.

On June 14, 1939, Memorial Hospital moved across the city to the new site. A mere 33 patients were in the old hospital, and they were moved. That was the easy part; the actual move filled 293 moving vans. At first, 199 beds were put into operation with plans for 250 more. The 199 beds were completely filled within a few months.

A million-volt x-ray machine was a new and improved unit, the product of the General Electric development team led by William D. Coolidge. With the exception of only three treatment machines brought over from the old hospital, all the therapy equipment was practically new, Dr. Herendeen wrote in a report.

The diagnostic section also received new equipment. For example: a new 500-milliampere rotating anode tube, a fluoroscopy unit that was almost new before the move, a new genitourinary x-ray table, a portable bedside unit, space for illumination boxes, and an expanded filing section.

The transfer of the 4 gm radium "pack" was made in a truck with Failla as a passenger and with police protection. Moving the additional 4 gm into the emanation plant required contracting with an out-of-state expert.

The radium facility was located adjacent to the x-ray division, with the physics group close by on the 2nd floor. Special wiring was run for radiology, and at the same time air conditioning was provided for most of radiology and for the operating rooms.

The therapy division had three radiologists, one fellow, and one technologist. The physics group had 17 people, still led by Failla. The group now included a new physicist, Leonidas D. Marinelli (1906–1974) and a new biophysicist, Robert S. Anderson, who succeeded Paul Henshaw. Marinelli, the inventor of the beaker that bore his name, conceived the idea of placing the sample in direct contact with the counting device. The beaker, for example, was used for analysis of I^{131} in liquids (mainly urine).

Meanwhile, the responsibility for the equipment, technical personnel, and logistics of roentgentherapy, in general, was Failla's. In addition, Failla acquired a valuable machine shop in the new hospital, occupying a space between shielded therapy rooms. There he installed the radium emanation extraction unit that had been moved intact from the old hospital.

In October 1939, the extended low-level radiation rooms were rebuilt on the eighth floor of the new hospital. These rooms were used for the Heublein unit, in which patients lived for two to three weeks in low-level radiation.

In 1937, among other duties, Failla served as section chairman of the physics program at the fifth International Congress of Radiology in Chicago. Quimby expanded her lecture on the physics of radiotherapy for physicians and published a syllabus that became a standard text. In later years, she was honored by the American Radium Society, the Radiological Society of North America, and the American Society of Therapeutic Radiation Oncology.

As for clinical radiotherapy, the medical staff adhered to the earlier concept, namely that surgeons of the various divisions were the arbiters of radium's use. Surgeons applied radium in the operating room and prescribed the courses of roentgentherapy carried out by their residents. At that time, the institution had no radiotherapist.

Cancer and World War II

On December 8, 1941, with the United States' entry into World War II, the armed forces had only about 2,000 physicians on active duty. The nation's medical schools were kept on an accelerated schedule, but male graduates in good health were taken directly into military service. Thus, soon after Pearl Harbor, Memorial physicians began to leave for active duty, placing strains on the reduced staff.

The number of experimental protocols dropped as those still at home were forced to emphasize patient care. Debate continued about the efficacy of preoperative radiation but, overall, cancer was still regarded as a surgical disease.

Now, however, Memorial set forth on a new dimension of radiation science when in 1941 internist John Leach and biologist Kanematsu Sugiura (1892–1974) obtained a small quantity of phosphorus 32, an artificial isotope produced in the cyclotrons at the University of California. They used it to study the effect of high-energy x-rays on the hearts and lungs of experimental animals. In time, artificial isotopes would become important in both diagnostic imaging and as replacements for radium emanations. During the war years, however, radiology at Memorial changed little, as Memorial physicians and scientists struggled to cope with their growing workload.

At this time, diagnostic radiology consisted of plain x-ray films and fluoroscopy. Fluoroscopic images were important when observation of motion was needed, but the procedure was time-consuming and subjected radiologists, even wearing leaded aprons, to considerable radiation exposure.

More About Failla

During the War, Failla became a consultant to the Manhattan Project and to the Metallurgical Laboratory in Chicago. Since Failla had always been concerned with radiation safety, he naturally became deeply involved with protecting personnel from radiation hazards, while at the same time remaining active in the nuclear program.

With the coming of peace, Failla was finally able to return to his laboratory, while continuing as a consultant to the Atomic Energy Commission and related agencies. Failla and his family spent their summers in Woods Hole, Massachusetts, where he installed a dual-tube, high-intensity x-ray unit at the Marine Biological Laboratory and became a summer member of the staff.

In the next few years, two new topics became important. The first problem was the matter of recovery from radiation effects, the so-called "time factor." In the late 1940s and 1950, for example, the adoption of radioisotopes in laboratories, operating rooms, and departments of medicine put a great number of workers at risk and created new radiation protection problems. Many of these workers could not be expected to understand radiobiology, and Failla knew by experience that guidelines were likely to be ignored by neophytes. The permissible dose for total body exposure had been reduced to 300 milliroentgens a week and for exposure of hands, to 1500 milliroentgens per week.

The second problem was an attempt to bring doses for x-rays, gamma-rays, and possibly other forms of ionizing radiation under a common unit. Failla suggested that the dose in tissue should be based on the number of ion pairs produced per gram. Essentially this concept went back to the energy absorbed dose of 1921, but was now backed by the development of equipment to measure it. It seemed to many that before such a unit as the proposed "tissue roentgen" could be accepted, it would be necessary to know how to measure gamma rays in roentgens, a quest eventually managed by many eminent physicists, as well as Marinelli and Failla.

Failla Moves to Columbia University

On December 31, 1942, Failla resigned from the position he had held for 27 years at Memorial, but he remained as a consultant. At this time, Columbia University announced the establishment of a Laboratory of Radiological Research with Failla as it director of radiology (physics) at the College of Physicians and Surgeons. Quimby and Titus Carr Evans (1907–1975), the original editor of *Radiation Research*, joined him at the new laboratories.

Roosevelt Hospital in New York obtained a loan of 50 gm of radium, then worth then over $1 million, from the Belgian Union Minière. It was intended for tele-curie therapy use by Douglas Quick (1891–1960), then director of the hospital's department of radiotherapy. The need to provide housing for the source while managing its clinical use and providing personnel protection was a challenge ideally suited for Failla, who designed a unit with a special collimator and a mechanism to reduce the exposure of personnel between treatments.

Throughout his scientific career, Failla's inventive ingenuity fueled design ideas for the equipment needed to solve problems. His engineering training and his ability to secure good instrument makers for his shop contributed to a large number of developments. Although Failla patented only a few of his discoveries and instrument designs, he made most of his ingenious inventions freely available to anyone.

He granted exclusive rights to a commercial company to use, distribute, and market his patented inventions, among them, a vacuum tube with a "floating potential grid," a standard ionization resistance for measuring unidirectional electric currents, and a method to arrange the above to provide a differential ionization meter for the evaluation of radiation quality. Many other inventions, however, were never patented.

Failla's last work was concerned with radiation mutagenesis, the induction of cancer by radiation, and finally a theory of aging based on the accumulation of somatic mutations.

In a paper presented at the RSNA in 1953, Failla postulated that although the daily exposure to radiation could be reduced, the accumulated total exposure was also important. Recovery was greater than irreversible injury in general, he wrote, but the reverse was true for highly specific ionization, such as that of alpha particles and neutrons, which can produce chromosome damage and somatic gene mutations. He further theorized that if the aging process is due to the accumulation of somatic mutations in all tissues of the body from various causes (heat, childhood diseases, heredity), then exposure to ionizing radiation simply increases the mutation rate and accelerates the aging process.

In 1960, Failla had reached the academic retirement age and decided to leave the city where he had lived for half a century and move to Chicago. He was offered a position as senior physicist emeritus by the Argonne National Laboratories, where his former student and associate, John E. Rose, was head of the radiological physics division.

Tragically, on a cold December morning in 1961 while being driven to work by John Rose, the two men had a head-on collision on an icy road. Rose was seriously injured and Failla died instantly, not a martyr to radiation as some others before him, but in an automobile accident.

Alfred Sloan, Charles Kettering, and the Sloan Kettering Institute

In 1942, the hospital had 213 beds and 571 employees. Much of the surrounding area was owned by the Rockefeller family, whose members contributed several other parcels of land and made several direct donations for research and education programs.

In 1945, with the war still going on, two major General Motors industrialists, Alfred P. Sloan, Jr. and Charles F. Kettering, announced that they would provide support for the creation of the Sloan-Kettering Institute in connection with Memorial Hospital. They believed that significant improvement in cancer treatment would have to come from research about the nature of cancer, whereas most of Memorial's research programs were at the clinical level.

With this pledge, a building was to be constructed across 68th St. from the hospital, with a primary commitment to basic research. In 1945, none of the later federal research-support programs existed. Thus the Sloan-Kettering Institute became a pace-setter for a new approach to cancer research.

With the end of the war, American institutions began to implement long-postponed plans and programs. Memorial began construction of the Sloan-Kettering Institute with orders placed for new equipment, including new resources for radiology.

Now members of the Memorial staff returned from military service, while many newly fledged young physicians were ready to undertake specialty training. At the same time, long-serving physicians could consider retirement, including Ralph Herendeen, who had been director of radiology since 1918.

New Leaders

After Ewing's retirement in 1940, the institution was directed by Cornelius P. Rhoads (1870–1959), who was primarily interested in laboratory research. With the departure of Failla, Marinelli became chief physicist at Memorial and head of physics and biophysics at the Sloan-Kettering Institute.

Then in 1948, Marinelli left to become the associate director and later director of radiological physics at Argonne National Laboratory in Chicago and, still later, Director of Biological and Medical Research. He retired in 1971.

John S. Laughlin:
A Founder of Medical Physics

John Seth Laughlin (1918–2004) was born in Canton, Missouri. He received his B.A. from Willamette University in 1940, an M.S. from Haverford College in 1942, and his Ph.D. in nuclear physics from the University of Illinois in 1947. At the start of his career at the University of Illinois, Laughlin conducted research on particle accelerators, particularly on the early cyclotrons with P. Gerald Kruger and Donald Kerst.

John S. Laughlin

In 1940, Kerst developed the betatron, a high-energy device used to accelerate beta particles (electrons) in a circular orbit. In addition to providing an increasing magnetic flux to induce electrons to accelerate and acquire energy, Kerst shaped the pole faces to establish a stable equilibrium orbit, with the electrons injected tangentially near this orbit. The accelerated electrons had a relatively monoenergetic spectrum, and their energy was easy to control. Kerst's first betatron operated at 2.3 MeV, the second at 20 MeV, and the next at 300 MeV.

During this period at the University of Illinois, Laughlin helped with the installation of the betatron and was involved in the first therapeutic applications of high-energy x-rays using a 20 MeV betatron, originally devoted to physics. In 1948, a graduate student at the University of Illinois developed a glioblastoma, and it was decided to use localized irradiation following surgery. Radiologist Henry Quastler and physicist Donald Kerst carried out a treatment with high-energy x-rays using 20 to 30 fields all angled and coming in from different directions, perhaps a precursor of conformal radiation therapy developed almost half a century later.

Although the dose was tumoricidal out to the margins of the lesion, the patient eventually died. A postmortem revealed no viable neoplastic cells in the irradiated region. This first use of high-energy photons pioneered the medical use of the betatron.

The Allis-Chalmers Manufacturing Company developed a commercial version of the betatron for reliable medical use. In 1948, one of these devices, a 24 MeV betatron, was installed at the Saskatoon cancer clinic in Canada under the direction of Dr. Harold E. Johns. Another device was installed at the University of Illinois, and nine months later in March 1949. In 1950, Dr. R. S. Harvey, chairman of radiology at Illinois, John Laughlin, and Lewis Haas, another graduate student, also at Illinois, started treating a relatively small number of patients using the betatron's high-energy electron beam.

Dr. James J. Nickson became chairman of Radiation Therapy at Memorial in 1950 and recruited John Laughlin to Memorial in 1951. There, he oversaw the installation of a 24-MeV betatron, donated by the Kress Foundation, and devoted to medical use. Its installation, completed in 1953, was the first high-energy electron-beam machine to be installed in a U.S. cancer center. A variety of high-energy accelerators were used almost exclusively for nuclear physics research; the betatron turned out to have practical applications in medical therapy and diagnosis. The absorption of electrons was shown to be quite different from the absorption of x-rays. As a consequence, the energy dissipated by the electrons was fairly uniform along the beam and terminated at a finite range. In a 1953 article published in *Electronics,* Laughlin wrote that low-energy x-ray beams produce their maximum dose on the skin, do not penetrate to great depths in tissue, are not well defined because of scatter, and are particularly destructive to bone. By contrast, he explained, the high-energy x-ray beam of the betatron is effective in irradiating deep-seated tumors with minimum damage to surrounding tissue. The maximum radiation dose is produced 3 cm to 5 cm below the patient's skin, and its side scatter is negligible. At these energies the beam is well-defined, permitting effective localization of the dose.

When the betatron's high-energy electrons are used, bone does not absorb any more dose per unit mass than soft tissue, circumventing a serious limitation in lower-voltage x-ray therapy. The maximum depth of penetration is proportional to the incident energy, while the decrease near the end of the penetration is fairly steep. Thus, the healthy tissue at greater depth is not exposed. Furthermore, Laughlin wrote, the betatron's electron beam could be used to treat lesions near the surface with finite maximum depth of penetration proportional to the incident energy.

The original betatron installed at Memorial was donated to the Smithsonian Institution in 1977.

In 1952, Laughlin was appointed chairman of the newly reconstituted Department of Medical Physics at the Memorial Hospital for Cancer and Allied Diseases and Chief of Biophysics at the Sloan-Kettering Institute. With the arrival of John Laughlin at Memorial, many gifted men and women were attracted to the growing staff. Medical physics experienced a period of growth and expansion with a brilliant flowering in an alliance between radiation physics and radiation therapy. Over approximately three decades, Laughlin's pioneering and legendary research eventually included every aspect of medical physics: advancing the quality of medical imaging and diagnostic radiology, improving the accuracy of radiation dosimetry and radiation safety, optimizing the delivery of radiation treatment, and studying the biological effects of radiation.

Laughlin made significant contributions to x-ray dosimetry and nuclear medicine. A few examples include an early interest in bone marrow dosimetry, the construction of a dual-headed gamma camera, the introduction of computer analysis in imaging applications, the potential for total-body scanning, the installation of a cyclotron for the production of short-lived radionuclides, and the placement of the first PET camera at the hospital.

A strong advocate of teaching, Laughlin and his colleagues taught hundreds of residents, graduate students, and radiological physicists. Many of the leaders in the current field of medical physics throughout the country received training in the "Memorial Program" created in this period.

In the mid-1950s, Rosalyn Yalow, a nuclear physicist and a close friend of Laughlin, and Solomon Berson had developed a radioimmunoassay procedure for insulin, based on the principle of competitive binding by antibodies of natural and labeled hormones. The method became the basis for numerous assays in diagnostic and physiological research, and for this work Yalow received the Nobel Prize for Medicine in 1977.

A Cyclotron at the Sloan Kettering Institute

Although cyclotron-produced radionuclides were used for biomedical research almost from the time that Ernest Lawrence dedicated his 60-inch unit as a medical cyclotron, such use was uncommon. Then in 1965, cyclotrons were installed at the Washington University School of Medicine by Michel TerPogossian and also at Massachusetts General Hospital by Gordon Brownell, who used cyclotron-produced positron emitters for metabolic studies.

In 1967, Laughlin installed a cyclotron at Memorial Sloan-Kettering Cancer Center in the new Kettering laboratory building, which had opened in 1964. A prototype machine designed and built by the Cyclotron Corporation could accelerate helium-3 ions as well as protons, deuterons, and helium-4 ions. The short-lived positron emitters were eventually widely used in PET scanning. This cyclotron was removed in 1995 and a new cyclotron was installed in the Citigroup Biomedical Imaging Center (CBIC), a consortium collaboration between MSKCC and the Weill School of Medicine of Cornell University.

Radiation Oncology: Coming of Age

In building Memorial's fledgling radiation therapy department, the recruitment of Laughlin for the physics department and the fostering of brachytherapy (by Dr. Ulrich Henschke and subsequently by Dr. Basil Hilaris) were significant development led by Dr. James Nickson. With great foresight, Dr. Florence Chu wrote, Dr. J. Nickson also established a Radiobiology Laboratory and expanded the department by increasing the staff and its meager equipment to include a 2-MeV x-ray machine and a 24-MeV betatron.

Nickson's most difficult task was to try to achieve total independence for the Radiation Therapy department from the "cancer specialist" surgeons. Following Memorial's long-standing tradition, the surgeons (and internists) refused to surrender their radiotherapy privileges, even though the techniques had developed at a pace far beyond their grasp. Weary of the endless battle, Nickson resigned in 1965 and moved to the University of Chicago. His associate, Dr. Marvin Glicksman, also left for the pion project at the Los Alamos Laboratory. Dr. Ralph Phillips, who had previously taken leave for illness, returned eventually to become the Chairman of Radiation Therapy. Phillips served until his retirement in 1968 when Dr. Giulio D'Angio of Philadelphia was recruited to be department chairman.

D'Angio strengthened the training programs and created new treatment and research programs for pediatric and other cancers, and he studied the radiosensitizing effects of hyperthermia. D'Angio also established a total skin electron bean therapy (TSEB) protocol, using linear accelerator electrons to treat skin lesions, and put Dr. Lourdes Nisce in charge of the program. This treatment protocol required major support from Medical Physics, including the construction of whole-body chambers to monitor the doses, the tailoring of blocks to lower the dose to the lung, and precise dose calibration at different treatment distances. Like his predecessors, however, D'Angio struggled with the surgical orientation of Memorial's clinical programs. He resigned in 1976, returning to pediatric oncology practice in Philadelphia.

Dr. Florence Chu served as Chairman of the Department of Radiation Therapy, later renamed the Department of Radiation Oncology, from 1977–1984. Born in Shanghai, China, in 1918, Florence Chu came to Memorial in 1949 as a special fellow in radiation therapy and entered the Department of Radiation Therapy under Dr. Phillips. After 1984, Chu remained in the department for two more years.

Chu worked to upgrade the five treatment units and to replace these units one by one. With the strong support of the physics team led by Laughlin, there were further developments in brachytherapy and, in addition, a sea change in the treatment policy for early-stage breast cancer. Memorial was known for advocating radical and supraradical mastectomies, but with mounting clinical data, the surgeons reluctantly accepted the newer treatment of breast-conserving surgery (lumpectomies) followed by focal radiotherapy.

In 1984, Dr. Zvi Fuks, previously chairman of Radiation and Clinical Oncology at the Hebrew University–Hadassah Medical School in Israel, became chairman of the Radiation Oncology department. As a further indication of the change in emphasis, Dr. Samuel Hellman, previously at the Joint Center for Radiation Therapy of Harvard Medical School, was named Physician-in-Chief of Memorial Hospital, the first non-surgeon to hold this position.

In the late 1950s and early 1960s, brachytherapy techniques still used radium needles and radon seeds. Given the growing concern about the harmful effects associated with the use of radium, the use of brachytherapy faced mounting problems from physicians, nurses, and patients, as well as from state, city, and hospital radiation safety regulations. Many hospitals, including Memorial, found it increasingly difficult to recruit and retain personnel willing to care for patients using high-energy radioactive materials.

If someone wanted training in modern brachytherapy, according to Dr. Basil Hilaris, author of a detailed book on the history of brachytherapy, the only place available in the late 1950s and early 1960s was Memorial Hospital. There, Dr. Ulrich Henschke and his group, including Dr. Hilaris, were developing innovative techniques based on afterloading and using iridium as a substitute for radium.

Henschke had arrived at Memorial Hospital in 1955, after working on afterloading at Ohio State with Dr. W. Myers, on small ^{60}Co and ^{198}Au radioactive sources. There he made practical the principle of afterloading radioactive sources for intracavitary or interstitial applicators.

Henschke remained at Memorial until 1967, at which point he left for Howard University where in 1970 he was appointed professor of radiotherapy. Upon Henschke's departure, Hilaris took over the clinical leadership in brachytherapy. In 1979, he was named chief of Memorial's newly formed Brachytherapy Service in the Department of Radiation Oncology, a position he held until 1988. Some two-thirds of all brachytherapy techniques used in the United States originated at Memorial, where over the years, the growth of the technology surrounding afterloading and the use of radium-substitute materials was a highly successful collaboration. Writing in 1989, Laughlin noted that in the 1960s these major innovations were important for using brachytherapy sources and in reducing radiation exposure for the staff.

The Era of Cobalt

Beginning in 1951 in the United States, reactor-produced cobalt-60 (^{60}Co) was advocated for several years for both externally and internally administered radiation treatment. The first commercial ^{60}Co teletherapy machines ("tele" from the Greek means *at a long distance*) became available in the 1960s. Harold Johns and his associates described the installation of such a machine in Canada, and the design was adapted for commercial purposes by Atomic Energy of Canada, Ltd.

Because ^{60}Co had the major advantage of not requiring associated power supplies, complicated acceleration apparatus, and had an energy just high enough to provide some skin sparing. Although linear accelerators eventually became the primary method for radiation treatment, cobalt units had a major advantage in that they could be installed almost anywhere.

The production of ^{60}Co permitted very high activities, leading to higher dose rates than the 250 kVp x-ray units. In addition, the availability of ^{60}Co spelled the end of radium teletherapy, along with the need to dispose of the seminal element and dismantle its housing.

The cobalt source was contained in a sealed capsule placed in a treatment head containing depleted uranium shielding and with trimmers to refine the treatment field. The source head was mounted on a gantry capable of rotating a full 360 degrees with the patient under treatment at the center of the rotation.

Because ^{60}Co decays with a half-life of 5.25 years, the sources had to be replaced periodically to maintain the desired treatment dose rate, typically from 100 to 150 rad/min. The source drawer moved from the treatment position to a shielded position by means of a hydraulic system. Since the source could never be turned "off," there would be some measurable radiation levels, so-called "leakage" radiation, at all times. The higher photon energies also meant that a more heavily shielded room was required than that needed for the 250 kVp x-ray units. Moreover, the entire facility required more space.

1967: New Department—the Firestone Center

A new Radiation Therapy department was planned in the 1960s where newer machines would be installed adjacent to the existing betatron. Planning and construction took place from 1965 to 1967, at which point the new department opened as the Russell Firestone Radiation Therapy Center. The complement of machines included the existing betatron, three ^{60}Co units, and one of the first commercially available linear accelerators, a Clinac 6 manufactured by Varian Associates, Inc. The department also used the 250 kVp machines for specific treatments until their removal after the construction of a new outpatient building in 1972.

In 1968, the James Ewing Hospital, built in 1947 by New York City for indigent patients and staffed by Memorial Hospital physicians, was merged with Memorial Hospital and renamed the Ewing Pavilion. The renovated division functioned initially as a hospital and was subsequently renovated in 1982 as the Schwartz Laboratory building, housing clinical and research laboratories and clinical offices.

The Arrival of Linear Accelerators

Before and during World War II, in England and the United States, oscillator tubes capable of relatively high power output at microwave frequencies were developed for the military and applied to radar. After the war, groups at Stanford University and in England led the development of linear accelerators designed for medical use.

In the decade following World War II, with the development of particle accelerators—especially the betatron and the linear accelerator—a new era arrived with the potential for optimal radiation treatment with high-energy x-rays and electrons. According to Laughlin, by 1982, in the United States alone, hospitals were using approximately 700 linear accelerators and 35 betatrons for cancer therapy.

The first medical linear accelerator was installed at the Hammersmith Hospital in England in 1953. In 1956 in the United States, Henry Kaplan used a Varian 6 MV linear accelerator to treat patients at Stanford University School of Medicine. Varian then manufactured a commercial 6 MV linac with a 360-degree rotation that was installed at UCLA in 1962. At Memorial, the first linear accelerator was installed in 1966 during the renovation of the department described above.

Treatment Planning and Dosimetry

The advance described above was accompanied by dosimetry problems. No longer could skin reddening (erythema) be a treatment guide because the maximum deposition of energy now occurred below the surface for deep-seated lesions. With the ability to concentrate the radiation dose, it became critical to locate the target region as accurately as possible, to plan the treatment in three dimensions, and to deliver the treatment precisely. The attending dosimetry problems were far more difficult than those for orthovoltage x-rays. Thus, treatment planning became an important aspect of radiation treatment. The calculation of the energy deposited per unit mass of tissue for x-rays, electrons, and heavier particles preoccupied many radiological physicists and, according to Laughlin, the publication of studies could "undoubtedly be measured in tons."

In the mid-1950s, Laughlin and his colleagues at Memorial Sloan-Kettering developed an absorbed-dose calorimeter using a thermally isolated wafer surrounded by a homogeneous absorbing medium. Both temperatures were monitored with thermistors that were part of a bridge. The unit, fabricated of carbon and polystyrene for the dosimetry of x-ray and electron beams, was calibrated electrically by passage of a known current through the wafer.

To achieve a tumoricidal dose and yet avoid irreparable damage to irradiated healthy tissue, the radiation physician had to achieve a specified tumor dose, usually within 5% uncertainty. Failure to adequately accommodate inhomogeneities in treatment planning could substantially affect the actual tumor dose, Laughlin wrote, while an error in delivering any beam of radiation on any day of treatment could negate care in the remaining treatments. Finally, Laughlin noted, the therapy plan could require modification during treatment if check-up examinations revealed changes in the tumor's configuration or its surrounding tissue.

Brachytherapy was highly dependent on computer technology. In the late 1950s, Laughlin and his physics colleagues were instrumental in using computers to calculate the dose distributions for implanted radionuclides. An IBM 1800 computer, then considered state-of-the-art, was installed in the cellar of the Kettering Laboratory building to perform treatment planning calculations and radionuclide imaging.

K. C. Tsien, a medical physicist, had previously demonstrated the feasibility of external beam isodose computation, while other members of Laughlin's group, particularly Mary Lou Meurk and Richard Nelson, developed computational methods for interstitial and intracavitary brachytherapy. The methods for radium dosimetry relied on calculated doses to specific locations (Points A and B). The system developed by Quimby defined doses from line sources, and it was eventually replaced by the Paterson-Parker system, which allowed for volume calculations. Treatment times were specified in terms of the equivalent times previously used with radium, mg-hours radium equivalent. It was becoming increasingly evident that the dosimetry calculations were going to require more complex methods.

Automatic computation methods developed at Memorial, MD Anderson, and other institutions, provided the dose distribution throughout the volume of the implant rather than at a few points. For example, a method developed in 1976 made it possible to determine the optimum strength and location of cesium-137 sources in intracavitary applications. At 16 specific points, the clinician could determine the necessary source strength at each source position.

Simulators Arrive

Dosimetry experiments with betatrons had made it clear that since the body was not uniformly dense, corrections to the treatment plan were necessary for inhomogeneities within the treatment field. It followed that similar problems existed for other radiation modalities.

Simulators developed from modest origins, in which diagnostic x-ray tubes were mounted on improvised stands connected to, or adjacent to, a treatment couch. In one of the first attempts to measure densities in cross-sectional anatomy, Laughlin and his associates mounted a sealed source on an arm that passed transversely over the body section and detected the radiation that passed through the body.

Toshiba adapted this methodology, substituting an x-ray machine for the source, and the company installed a transverse axial tomograph in the department in the late 1960s. The results were used to modify treatment plans to account for density differences and anatomy. It soon became apparent that more accurate plans would be possible with more information.

The first commercial simulator was installed in the 1970s, using an x-ray tube as the radiation source. Simulators had the same geometry as that of the treatment machine and allowed a simulation of the patient's treatment prior to delivering the first treatment on a radiation machine. During the simulation sessions, kV x-ray images of the patient's treatment field were captured on film and reviewed by the physician. Thus, with the use of a simulator, refinements in the treatment field and patient geometry could be incorporated before the start of treatment. During patient setup for treatment, the simulation films may be compared to portal films to verify the accuracy of the radiation field to be delivered.

Radium Bows Out: Act I

With the availability of other radioisotopes—such as ^{60}Co, ^{137}Cs, ^{192}Ir, and ^{125}I—the days of utilizing radium were coming to an end. With time, technologies were developed to encapsulate these radioisotopes in tubes, needles, or seeds to replace radium needles and radon seeds in intracavitary and interstitial implants.

In addition, by 1970, the blown-soft-glass in the radon plant had suffered radiation damage. Replacement of the glass would be a difficult and dangerous operation. Another extreme taxing problem was how to replace the flasks containing the radium salts dissolved in acid. Furthermore, by now only one or two thoracic surgeons continued to order radon seeds, which were implanted into the lung using traditional methods by the same surgeons. A critical decision was then made to discontinue the radon plant.

The above decision left Laughlin and Jean St. Germain, head of the Radiation Protection section, with the difficult task of disposing of Memorial's radon plant, its contaminated equipment, contaminated metals, and the decontamination of the areas that housed the plant.

Originally, four grams of radium salt had been dissolved in acid and used in the radon plant. Between 1944 and 1948, a substantial portion of the solution had been removed, dried, and made into sealed tubes and needles. This left 1.76 grams in solution in the radon plant. Because the EPA accepted only sealed sources, the agency was never an option for disposal, according to St. Germain, who described the undertaking.

Before dismantling the plant, however, permission for disposal was needed from each of the parties who had donated the radium. The hospital administrators at that time were Donald Teece and Glenn Wesselman. Theirs was the task of obtaining permission from the Douglas family (a member of the Douglas family always had a seat on Memorial's Board of Managers), the U.S. Department of the Interior, Rockefeller University, and the trustees of Cornell University.

The custodian of the plant and the sealed radium sources was Thomas Nicholson, who had begun work at Memorial in the early 1940s. He regularly pumped the plant, produced and calibrated the radon seeds, and knew exactly how each stage functioned. The first and critical stage was the removal of the flasks containing the radium salts. A commercial firm was hired to assist in the disposal of the plant, and the firm prepared shielded drums specifically designed to accept the flasks and convey them unbroken to a designated burial ground.

Nicholson, George Magyar, a member of the radiation safety staff, and St. Germain pumped down the plant for the final time. They calculated that they had about 25 minutes before activity would build up to a significant level. Using a hot glass rod, the flasks were broken off the plant and were placed in the casks to be shipped out. The casks were put on a waiting truck and began their journey to the designated site.

Over the next few months, the various stages of the plant were dismantled and shipped to designated locations. These shipments included all the contaminated glass and mercury—including the gold tubing that had contained the radon gas—other metal parts, as well as the safe in which the flasks had been contained. There remained some unused 24 kt gold tubing. This gold was uncontaminated and was eventually sold in 2011.

The decontamination of the areas housing the plant, including walls and surfaces, took months. The removal and drying of the radium salts from the solutions in the plant had taken place in an area immediately outside the radon plant room, in what was most probably an inadequately ventilated area. This operation resulted in large amounts of contamination on all surfaces, including the ceiling. Removal of this contamination was attempted using various solvents and chelating agents.

Eventually, however, it was decided that the remaining activity was fixed and could not be further removed. The endpoint of the decay series was an isotope of lead with a 22-year half-life. The area of the retired plant remains a "controlled area," and it is likely to stay so for several generations, St. Germain wrote.

Radium Disposal, Act II

The sealed sources that had been fabricated from the original radium salts remained at Memorial until 1989. The sources were stored in a large safe, along with other sealed sources used at Memorial, as well as sources and random uranium ore samples donated to the hospital.

During the 1970s and 1980s, the cost of disposal had risen significantly. In addition, concerns about the disposal of low-level radioactive waste were raised in Congress and by those states that had previously served the entire nation. It became obvious in 1988 that the window for disposal of the sealed sources was closing, and something needed to be done. St. Germain wrote in a memo to Dr. Samuel Hellman, Physician-in-Chief of Memorial Hospital and Dr. Zvi Fuks, Chair of Radiation Oncology, recommending the disposal of all the content of the large safe.

In December 1989, the remaining radium sources, the various ores, and other sealed sources were delivered for disposal. As some of the sealed sources had leaked over time and had been sealed in glass, to open the safe and remove the contents, the safe was moved to a "radioactive" hood large enough to accommodate the safe and the persons who would remove the actual sources, and with sufficient air flow to dilute any potential release.

Special shielded drums were provided by a commercial vendor. All of the radium sources, totaling ~2.76 grams, and other radioactive materials were sealed in the drums and sent to a designated disposal site. Decontamination of the room where the safe had been housed was accomplished, although because of the large shielded walls, this area remains under radiation control and is now used for a newer generation of sealed sources.

It should be emphasized, however, that the purpose of the original Douglas donation had been fulfilled. Thousands of patients had been treated with radium, and much human suffering had been relieved. Now it was time to move to a new era.

Brachytherapy Development after Radium

Even before the removal of radium from Memorial, developments in brachytherapy were occurring with new radiation sources with various attractive attributes. ^{192}Ir was first used in 1958 and ^{125}I in 1965. These radiation sources made it possible to discontinue using radium and radon. Ulrich Henschke, M.D., and Basil Hilaris, M.D., began using these new radiation sources as they became available commercially, with the help of medical physicists Garrett Holt, Steve Balter, and Gian Ragazzoni.

In 1969, Lowell Anderson, Ph.D., was recruited by Laughlin to Memorial Hospital. Dr. Anderson would have a significant impact on the physics of brachytherapy for the next three decades. Dr. Anderson was born on September 3, 1930, in Spokane, Washington. He received his B.S. degree in Physics from Whitworth College in Spokane, WA, in 1953 and his Ph. D. (Biophysics) from the University of Rochester in Rochester, NY, in 1958. After an 11-year appointment as biophysicist at Argonne National Laboratory, where he worked mainly on developing instrumentation for neutron dosimetry, he joined the Department of Medical Physics at MSKCC.

Shortly after his arrival, because of his neutron dosimetry background he was selected to coordinate a contract with the Department of Energy to evaluate the use of ^{252}Cf neutron sources in interstitial brachytherapy. That project led to his long-time interest and specialization in brachytherapy physics. For the ^{252}Cf project, he and his colleagues had to establish the dose pattern of individual sources, not only from neutrons, but also from gamma rays, so that there were two doses to be computed each time, and an RBE assumed for the neutron dose. Because of the complexity of the dosimetry and the concern about radiation safety, the project lasted for only a couple of years.

Even before Lowell's arrival in New York, ^{192}Ir was used for temporary implants and ^{125}I for permanent implants at MSKCC. ^{125}I permanent implants were first performed for head/neck nodes in 1965, for prostate cancer in 1967, and subsequently for lung cancer, with seeds provided by the Lawrence Soft-Ray Company. As quantitative imaging for tumor volume determination, prior to surgery, was not available, medical physicists were called during surgery to provide support if radioactive implantation was to be performed. The average dimension method, an extension of the technique for radon seeds, was used to determine the amount of ^{125}I radioactivity to be implanted. The average dimension method is to measure the three dimensions of the tumor and multiply the average dimension by 10 to obtain the number of millicuries for radon and by 5 for ^{125}I. That method preceded Dr. Anderson's arrival at Memorial.

Pursuant to the ^{252}Cf project, Lowell was appointed head of Brachytherapy Physics, and he began to make significant contributions. One early example was the concept of matched peripheral dose, instead of minimal peripheral dose, for an improved quantitative index of the quality of the implant. A very important contribution was his creation of a ^{125}I implant nomograph for a rapid calculation of the seed spacing required for a uniform distribution of the number of seeds specified by the dimension averaging method. The nomograph performed the spacing calculation based on a power function of average dimension, which is well beyond the reach of mental arithmetic. This concept was subsequently adapted and refined for planar implants, and then for permanent implants of ^{103}Pd seeds.

In dosimetric measurements, he performed or collaborated in determining the dose distributions for ^{137}Cs, ^{192}Ir, ^{125}I, ^{103}Pd, and ^{252}Cf sources. With the arrival of the computer era and high-dose-rate implantation techniques, Dr. Anderson developed least-squares optimization techniques for high-dose-rate (HDR) remote afterloading. He published a planning system for stereotactic temporary implants of brain tumors with high-strength ^{125}I seeds involving least-squares optimization of seed positions. He applied a variation of this system to surgical-deficit mold treatments.

Perhaps his most important seminal contribution was the idea of a "natural" ($-3/2$ power) dose-volume histogram to assist implant design and evaluation. He also wrote the software to calculate patient-specific "natural" ($-3/2$ power) dose-volume histogram. He directed the compilation of an extensive treatment-planning atlas for intraoperative HDR remote afterloading using quasi-planar, flexible-plastic applicators. Such applicators became known, collectively, as the HAM (Harrison Anderson Mick) applicator for intraoperative implants to recognize, as well, the contributions of the radiation oncologist, Louis Harrison, and the manufacturer, Felix Mick.

For nearly 20 years, Lowell served as the coordinator of a lecture course in medical radiation physics at MSKCC[*], and he was a lecturer in basic radiation physics and in brachytherapy physics. Participants in the course included MSK residents in radiation oncology and fellows in medical physics, as well as residents in radiology from Cornell University Medical College, where Lowell held an academic appointment. He also supervised a number of degree candidates and post-doctoral fellows in various brachytherapy physics research projects.

*C. Clifton Ling was taught by Dr. Anderson in 1972.

Educational Programs and the
J. S. Laughlin Visiting Professorship

The educational courses for the Department of Medical Physics included lecture courses for residents and laboratory-intensive courses for residents and for medical physics post-doctoral trainees.

During the late 1960s and early 1970s, lectures on various aspects of radiological physics, coordinated by John Laughlin and then Lowell Anderson, were presented twice weekly to first-year residents in radiology, radiation oncology, and nuclear medicine. It should be noted that the division of the various fields of radiology and the requirements for the American Board of Radiology (ABR) certification in these fields were not as distinct at this time as they would become later. Medical physics trainees also attended the lectures. Members of the department who taught various sections of the course included Jean St. Germain, Lawrence Rothenberg, Larry Simpson, Gerald Kutcher, Garrett Holt, Malcolm Powell, Martin Graham, Keith Pentlow, Gerald Randall, and Stephen Balter.

In the early 1990s, when the intensive course format was no longer feasible for the radiation oncology residents due to their clinical duties, a new two-year-long, once-a-week lecture course was established, managed initially by Jerry Kutcher, and subsequently by Howard Amols and James Mechalachos. Many members of the department are still involved in teaching the course with an updated curriculum, including topics such as 3D conformal radiation therapy (3DCRT), intensity-modulated radiation therapy (IMRT), and later image-guided radiation therapy (IGRT). Marco Zaider and Gil Cohen taught the subjects related to brachytherapy.

Laboratory and Intensive Courses for Residents

An intensive multiweek course for radiology and radiation oncology residents was also provided during their third year to prepare them for the American Board of Radiology (ABR) Certification Examinations. Initially, prior to the split between diagnostic radiology and radiation oncology, a six-week intensive course was offered, often referred to as the "Laboratory Course" because students undertook laboratory activities, e.g., measurements in radiographic and fluoroscopic x-ray rooms, calculations for radiation treatment planning and brachytherapy seed placement, and radiation protection measurements of shielding barriers. At first, all the lab courses were coordinated by Gian Ragazzoni. Following Ragazzoni's retirement, Larry Rothenberg took over the Diagnostic Radiology Lab course, and more recently, Douglas Ballon, who holds joint appointments at MSKCC and Weill Cornell Medical Center. Most of the Diagnostic Radiology Residents were from New York Hospital (now Weill Cornell NY Presbyterian Medical Center). On occasion, residents from sister hospitals attended, including Hackensack Hospital, Lenox Hill Hospital, and the Bronx Veterans Administration Hospital.

When the ABR split Diagnostic and Therapeutic radiology into separate disciplines, the Laboratory Course also separated into two four-week intensive courses, respectively. Experimental measurements continued to be included in the Diagnostic Radiology course. Most of the Radiation Oncology residents were from MSKCC, but there were additional attendees from other area medical centers, such as Bronx VA, St. Vincents Hospital, and Lenox Hill Hospital. In addition, a third intensive course was added for Nuclear Medicine residents, conducted by G. Ragazzoni, M. Graham, and K. Pentlow.

Among the physicians trained in these courses, many became leaders in their departments, e.g., radiologists Robert Min, Chair of Radiology at Weill Cornell-NYP, Lawrence Schwartz, Chair of Radiology at Columbia-NYP, radiation oncologists Beryl McCormick, past Acting-Chair and current clinical director at MSKCC, Dattatreyudu Nori, Radiation Oncology Chair at NYH Queens, and Bhadrasain Vikram of NCI and former Chair at Montefiore.

Starting in about 1980, with the development of many new types of equipment and techniques for both imaging and therapy, the course became more of an intensive lecture course, with no time available for laboratory experiments. For the Imaging Course, computed tomography, magnetic resonance imaging, molecular imaging, and many other new topics were later incorporated into the curriculum.

Physics trainees were required to take the lecture and the laboratory courses in their specialty area alongside the medical residents. This requirement resulted not only in many lasting collegial relationships, but it had the benefit of imparting knowledge of each other's discipline in a challenging environment for both groups.

For logistical reasons, the Radiation Oncology Lab Course ended in 1988 and the Nuclear Medicine Course in 1990. The Diagnostic Imaging Lab course ended in 2004.

Research and Clinical Training Programs for Medical Physics Postdoctoral Fellows and Residents

The Postdoctoral Fellowship program in the Department of Medical Physics has a rich history going back to at least the late 1960s. It was designed to provide physicists with a combination of formal education and intensive on-the-job training not generally available in most institutions. The focus of the program was on the quantitative uses of radiation for diagnostic and therapeutic purposes in order to qualify the fellows for professional responsibilities in clinical settings.

Starting in 1969, the Department under John Laughlin received a grant from the American Cancer Society to train postdoctoral medical physicists in both research and clinical aspects of imaging and radiation therapy physics. Initially, a one-year stipend was awarded, later extended to two years. This training grant was continued until the early 1990s, when ACS discontinued that particular program.

In 1994, Ling obtained a NCI/NIH T32 training grant entitled Training in Radiation Oncology Sciences (TROS). The TROS T32 supported four research trainees a year in various aspects of radiation oncology sciences, e.g., radiobiology, medical physics, etc. TROS received continued funding until 2012.

Over the past decades, a number of dedicated senior faculty members have coordinated the Postdoctoral Fellowship Program, including Gian Ragazzoni, Jason Koutcher, Douglas Ballon, and Lawrence Rothenberg. Their activities in this area included preparing clinical rotation schedules, one-on-one meetings with the fellows, organizing final oral examination committees, attending and participating in Center-wide GMEC (Graduate Medical Education Committee) meetings, and writing letters of recommendation to prospective employers, state and national regulatory and certification bodies, as well as providing teaching sessions and hands-on training for the fellows in their own areas of expertise. The evolution of the certification program of the ABR into sub-specialties was reflected in a similar diversification among the various specialties within Medical Physics. Therefore, medical physics trainees were require to elect training either in Imaging Physics or in Radiotherapy Physics.

Although this program provided excellent training, the teaching faculty was interested in having the trainees participate in the research program. During the early 1990s under Ling's direction, the faculty of the Medical

Physics Department greatly expanded its research activities. This expansion led to increased funding for postdoctoral fellows who could contribute to the research effort. To have the fellows participate in research and still receive significant clinical experience, the program was expanded to three years, with the first portion dedicated primarily to research and course work and the remaining portion allocated mainly to intensive clinical training.

During the late 2010s, with the impending ABR requirement for candidates to have completed a CAMPEP-approved medical physics residency program, the so-called "2+2" program was designed. This program was to provide two years of research as well as the required two years of intensive clinical training. Requisite didactic coursework, if needed to remedy the Fellow's background, is taken in parallel with research work during the first two years. The didactic courses could include radiobiology, anatomy and physiology, and three to four ABR-required medical physics courses. The belief is that this program will continue to attract individuals with a diverse spectrum of modern scientific expertise, thus continuing the flow of ideas that has enriched this field since its earliest days.

With the "2+2" program in place, the Department applied for accreditation by the Commission on Accreditation of Medical Physics Education Programs (CAMPEP) and received full accreditation in 2009, with Doracy Fontenla as the program director and Lawrence Rothenberg as the associate director.

John S. Laughlin Visiting Professorships

A highlight of the year for the Department of Medical Physics is the John S. Laughlin visiting professorship. This annual event was established through a generous gift from Dr. Helen Quincy Woodard, a long-time faculty member at Memorial Sloan-Kettering. The professorship recognizes the distinguished career of Dr. John Laughlin, Chairman of Medical Physics at MSK from 1952 to 1988, spanning almost four decades. The JSL visiting professors from 1990 to 2012 included:

1990	Dr. Jean Dutreix and Dr. Andrée Dutreix
1991	Dr. Edward R. Epp
1992	Dr. Michel M. Ter-Pogossian
1993	Dr. John R. Cameron
1994	Dr. Rosalyn Yalow
1995	Dr. Thomas Wheldon
1996	Dr. Samuel Hellman
1997	Dr. Samuel J. Dwyer, III
1998	Dr. David Bragg and Dr. Robert Sutherland

1999   Dr. Thomas Rockwell Mackie
2000   Dr. Eric J. Hall
2001   Dr. Steven Webb
2002   Dr. Albert J. Van der Kogel
2003   Dr. Jens Overgaard and Dr. Marie Overgaard
2004   Dr. R. Mark Henkelman
2005   Dr. Rakesh K. Jain
2006   Dr. Herman Suit
2007   Dr. Simon Cherry
2008   Dr. George T.Y. Chen
2009   Dr. Willi A. Kalender
2010   Dr. David Jaffray

The choice of the visiting professor is made by a department committee. Criteria include distinction and accomplishment in the field, and the ability to contribute to programs that were being developed at the time in the Department. The visiting professor usually spends two to three days at the Center to interact with the professional staff and trainees.

For example, during Dr. Edward Epp's visit, he chaired an extramural committee to review the 3D-CRT Program Project submission to the NCI. Michel Ter-Pogossian came specifically to advise on PET development, and Thomas Wheldon on the use of radiolabeled antibody cancer therapy. Samuel Hellman's visit focused on the department's developing program in IMRT. Extensive discussion for the clinical implementation of PACS was on the agenda for Samuel Dwyer's visit.

During the development of biological and hypoxia imaging, Drs. Robert Sutherland, Albert Van der Kogel, Marie and Jens Overgaard, Mark Henkelman, and Rakesh Jain participated. Herman Suit's visit coincided with the department's consideration of proton therapy, and Simon Cherry participated during the integration of PET/MRI imaging.

These examples, and the detailed agenda of Dwyer's visit (see below), show how the department went about developing its programs synergistically with the JSL visiting professorship:

PACS Workshop
Samuel J. Dwyer, III, Ph.D.
John S. Laughlin Visiting Professorship
Memorial Sloan-Kettering Cancer Center
May 29–30, 1997

MAY 29

9:00 a.m.   Visit: Dr. Clifton Ling, Dr. John Laughlin
9:45 a.m.   Visit: Dr. Ronald Castellino, Chair of Radiology, MSKCC
Radiology clinical areas: MedSpeak, PACS, RIS

Workshop I
10:30 a.m.   Introduction, Peter Kijewski
10:35 a.m.   Review of a PACS implementation, Sam Dwyer
11:05 a.m.   Integration of Radiology systems: PACS, RIS, and Voice
Recognition, Peter Kijewski
11:25 a.m.   End of session

12:00 p.m.   **Lunch** with postdoctoral fellows

Workshop II
1:30 p.m.   Disease Management System and PACS, Karen Malbin
1:45 p.m.   The impact of computer automation on Radiology workflow
and procedures, Lawrence Schwartz
2:15 p.m.   Discussion
2:30 p.m.   End of session

3:00 p.m.   **Workgroup Session I**
Analysis of cost and benefits of PACS
Sam Dwyer, Patricia Soto, Karen Malbin, Ronald Castellino,
Lawrence Schwartz, Robert Heelan, David Panicek

5:00 p.m.   **John S. Laughlin Lecture**

6:00 p.m.   **Reception**

7:00 p.m.   **RAMPS Dinner**

MAY 30

9:00 a.m.   **Visit to clinical areas:**
- Radiation Oncology treatment planning
- Surgery, Offices
- DMS

10:00 a.m. **Workgroup Session II**
Radiation Oncology and PACS
Clifton Ling, Zvi Fuks, G. Mageras, Jerry Kutcher, Chen Chui

10:45 a.m. **Workgroup Session III**
PACS quality assurance
Sam Dwyer, Lawrence Rothenberg, Chester Mah

12:00 p.m. **Lunch**

2:00 p.m.   **Workgroup Session IV**
The Virtual Hospital: Implementing PACS at multiple hospital
locations
Sam Dwyer, Lawrence Schwartz, Robert Heelan, Patricia Soto
Karen Malbin, Ronald Castellino, David Panicek

2:45 p.m.   **Workgroup Session V**
Reading stations for different modalities: CR, CT, MR,
Ultrasound, Nuclear Medicine
Sam Dwyer, Attending Radiologists

3:30 p.m.   **Workgroup Session VI**
Management topics
Sam Dwyer, Clifton Ling, Ronald Castellino, Lawrence
Schwartz, Robert Heelan, Patricia Soto, Wei-Tih Cheng

Radiation Protection

Early Works in Radiation Safety

The development of radiation protection principles and practices at Memorial Hospital, and at other centers, were necessitated by and coincided with the discovery and use of radium, radon, and x-rays. The use of radiation for diagnosis and treatment began at Memorial Hospital as early as the Ewing and Janeway era. As of 1933, Memorial Hospital possessed 4614 mg of radium, and radium's clinical usage required stringent radiation protection methodology. Failla and his colleagues had the dual role of developing both clinic physics applications, as well as radiation protection procedures. They pioneered the concept of 'optimization' in radiation protection, facilitating the beneficial use

Jean St. Germain

of radiation while minimizing risks. This concept continues to be the foundation of modern medical health physics. Edith Quimby was particularly concerned with radiation protection of both patients and staff. In addition to establishing the first formal personnel monitoring program using film badges, Failla and Quimby also employed the use of 'dose' films as standards for comparison. They subsequently developed the use of densitometers for reading film badges, improving accuracy, and establishing standardization.

Professional Radiation Protection Societies and Their Influence

As early as 1898, the Röntgen Society (currently the British Institute of Radiology) established a committee on x-ray injuries, thus initiating the discipline of radiation protection. In 1928, prominent radiologists and radiation scientists gathered at the second International Congress of Radiology.

During that Congress, Failla and Quimby were on working committees to develop an accepted set of definitions involving radioactivity and radiation exposure. These included the adoption of "roentgen" as a unit of x-ray exposure, which ultimately was applied for radiation protection purposes. In addition, they established the International X-ray and Radium Protection Committee, noting "...*the dangers of over-exposure to X-Rays and radium can be avoided by the provisions of adequate protection and suitable working conditions. It is the duty of those in charge of X-ray and radium departments to ensure such conditions for their personnel.*" This committee later evolved to be the International Commission on Radiological Protection, or ICRP.

In the United States, the Advisory Committee on X-ray and Radium Protection was formed in 1929 and included Lauriston S. Taylor of the Bureau of Standards, W. D. Coolidge of General Electric, and G. Failla of Memorial Hospital, amongst others. It was subsequently chartered by the U.S. Congress as the National Council on Radiation Protection and Measurements, or NCRP. Many members of the Medical Physics Department of Memorial Hospital were involved in these organizations. In 1934, Taylor, Rolph Sievert (Sweden), and others in the ICRP wrote: "...*the Commission would welcome information from those having experience in [telecurie] radium treatment.*" Memorial Hospital provided such experience and more. In 1941, the National Bureau of Standards issues a 15-page handbook which provided general measures for the safe handling of radioluminous materials, including a permissible residual body burden for radium.

The term "health physics," synonymous as the field of radiation protection, is believed to have had its origin at the University of Chicago in 1942. At the time, Robert Stone of the Health Division, and Arthur Compton of the Metallurgical Laboratory, were part of the Manhattan Project (1939–1946). During the same time period, medical physicists in the New York metropolitan area began to compare instrumentation and measurements of the amount of radioactivity in medical administrations. These efforts led to the "New York Millicurie," which was vital in providing uniformity and accuracy, as no national standard existed at the time. By 1948, regular meetings were held under the aegis of the Radiological and Medical Physics Society (RAMPS), with radiation protection as one of its mandates. The constitution of RAMPS was not written until 1954 and was co-authored by Rosalyn Yalow and John Laughlin. RAMPS is now the New York Chapter of the AAPM (American Association of Physicists in Medicine).

After World War II, x-ray units with higher energies and new radioisotopes became available. These necessitated the updating of existing radiation protection guidelines. In 1954, Failla, Quimby, Marinelli, and Laughlin of Memorial Hospital were all involved in the ICRP and its committees. Updated were recommendations on the permissible dose for external and internal radiation, protection for less than 2 MV x-rays, protection for greater than 2 MV x-rays, for beta and gamma rays, for particles (alpha, neutrons, protons), and procedures for the handling of radioactive isotopes and the disposal of radioactive waste.

The United Nations Scientific Committee on the Effects of Atomic Radiation (UNSCEAR) was established in 1955 during the General Assembly of the United Nations. Its mandate was to assess the levels of radiation exposure and their effects, and report back to the United Nations. In 1958, UNSCEAR issued the first scientific report on the effects of atomic radiation. The first report included input from Laughlin and other New York physicists, Failla, Harley, and Eisenbud. Since then, UNSCEAR continues to fulfill its mandate with subsequent reports, and governments throughout the world rely on these as the basis for evaluating radiation risk and for establishing radiation protection measures.

In 1955, at a Health Physics Conference at Ohio State University, Frank Bradley suggested, and seconded by Elda Anderson of the Oak Ridge National Lab, the formation of the Health Physics Society (HPS). Laughlin of Memorial Hospital served as its president from 1960–1961, during which time the American Board of Health Physics (ABHP) was incorporated to support the certification of professional health physicists. Many Memorial Hospital staff, including Jean St. Germain, Lawrence Rothenberg, Lawrence Dauer, Matthew Williamson, Brian Quinn, Bae Chu, Daniel Chiappetta, Daniel Miowdownik, and others have served or continue to participate with the Medical Health Physics Section of the HPS. Several of them served several terms on the Executive Board of the local Greater New York Chapter of the HPS (GNYCHPS) which began in 1959. Over the years, the Memorial Hospital team has supported multiple GNYCHPS scientific meetings. In addition, Memorial Hospital hosts the annual ABHP certification examination, with Dauer and Williamson serving as proctors. Also, RAMPS and GNYCHPS annually selects and honors the "Failla Lecturer." Several Memorial Hospital staff members have received that honor.

The "Atoms-for-Peace" Program and the "Radioisotope Revolution"

Although cyclotrons were able to generate certain radioactive isotopes (e.g., ^{130}I and ^{131}I) in limited quantities, nuclear reactors became available after World War II to produce large quantities of man-made radioactive elements. In 1949, Failla noted that "...*artificially produced radioactive isotopes are distributed by the Atomic Energy Commission according to strict rules on account of the dangers inherent in the handling of such materials. Before an individual can obtain radioisotopes, he has to present evidence that he possesses the knowledge, facilities and equipment to meet the minimum standards of safe handling.*" Such development was enhanced beginning in 1953 with the "Atoms-for-Peace" program. These new isotopes—which included ^{32}P, ^{60}Co, ^{137}Cs, ^{192}Ir, ^{125}I, ^{103}Pd, and ^{131}I—were much safer to handle and administer than radium or radon seeds. Indeed, a revolution in imaging and treatment began with these isotopes, e.g., the use of ^{60}Co in teletherapy, ^{192}Ir in low- and high-dose-rate brachytherapy, and ^{131}I in imaging of thyroid cancer.

In 1967, collaborating with the U.S. Department of Energy (DOE) and DOE National Laboratories, Laughlin installed a cyclotron (Model CS-15, Cyclotron Corp.) in the Kettering Building of the Sloan Kettering Institute. This was one of the earliest hospital-based cyclotrons for the production of novel radionuclides for imaging and therapy. Subsequently, modern cyclotrons were acquired. In collaboration with Cornell Medical College, a dual-beam unit (Model TR19/9, Ebco Technologies) with both liquid and solid targets was installed at the Citicorp Imaging Center in the early 2000s. More recently, the Radiochemistry and Cyclotron Facility brought into operation the General Electric PETtrace 880 cyclotron, which routinely produces positron-emitting and photon-emitting radionuclides for the synthesis of radiotracers in medical research and application. Medical health physicists support these efforts in many aspects, e.g., in the design of the facility relative to radiation protection, in establishing radiation safety procedures in radionuclide production and handling, and in the use of radioactive reagents in medical research and patient care.

More than Half a Century of Progress in Radiation Protection

An incredible era of medical physics and radiation protection began in the early 1950s that could be termed the 'volts to vaults' era. In 1952, Memorial Hospital recruited John S. Laughlin as its new chairman of the department of Medical Physics. In addition to conducting seminal work on the use of electrons from betatrons for the treatment of cancer, he led original research in calorimetry, radiation dosimetry, radiation protection, and radiation biology. He served important roles in several national organizations that included radiation protection in their mandates. Specifically, he was the president of the Health Physics Society from 1960–1961, a founding member of AAPM, and served as its president in 1964. In 1954, he contributed to the NCRP Report No. 14 on radiation protection of betatron-synchrotron radiation of 100 MV. In 1970, Laughlin chaired the writing group of the ground-breaking NCRP Report No. 37 on the precautions in the management of patients who have received therapeutic amounts of radionuclides. In 1970, Lowell Andersen co-authored the NCRP Report No. 38 on the protection against neutron radiation. This was followed in 1972 by NCRP Report No. 40 on radiation protection related to brachytherapy sources, with Edith Quimby as a contributor. And in 1981, Laughlin participated in NCRP Report No. 69 on dosimetry of x-ray and gamma-ray beams for radiation therapy.

One of Laughlin's most important decisions with regard to radiation protection at Memorial Hospital was the recruitment of Jean St. Germain. Over the years, St. Germain became the very definition of "radiation protection" in patient care and medical research. Following her completion of graduate study at Rutgers University and a fellowship at Brookhaven National Laboratory, St. Germain was appointed, in November 1967, as a Fellow in the Department of Medical Physics, under the tutelage of Laughlin and Garrett Holt. At the end of her fellowship, she was appointed to the faculty and steadily rose in rank to Attending Physicist. She served as Assistant Radiation Safety Officer, and then Radiation Safety Officer, for 50 years, guiding and presiding over the incredible growth of the institution. She also served as an interim chair of the Department of Medical Physics from 2007 to 2010 and subsequently as a Vice Chair for Clinical and Educational Affairs. Her contributions to the fields of medical physics and health physics were vast and significant. She held leadership roles in several professional societies. In addition, St. Germain served as a member of New York State advisory committees on medical and radiological health, and as a special examiner for the New York State Civil Service Commission.

During these decades, the use of radioactive material and radiation at Memorial Hospital were governed by a radiation safety program under the authority of several dozen licenses, registrations, and permits with local, state, and federal regulating bodies. At the main campus, most work is performed under large, broad-scope licenses allowing research/development and clinical applications on both human and laboratory animals.

In education programs, Memorial staff instructed health physics and radiation safety concepts to generations of medical physicists, radiologists, radiation oncologists, and other professionals, both at Memorial Hospital and throughout the New York metropolitan area. In the design of radiation facilities, members of the medical health physics section have participated in designing hundreds of treatment vaults throughout New York City, New York State, and New Jersey. Some of these had unique shielding designs, which have been adopted nationally and worldwide. These vaults have housed betatrons, ^{60}Co teletherapy units, linear accelerators, and high-dose-rate remote afterloaders.

Our Medical Physics staff continues to be influential in setting radiation protection standards and guidance throughout the decades leading up to the new millennium. In 1989, St. Germain was a member of the Scientific Committee (SC) for the NCRP that produced Report No. 105 on radiation protection of medical and allied health personnel.

More recent efforts of medical health physicists included clinical radiation safety for the use of ^{192}Ir, ^{125}I, and ^{103}Pd; the storage and disposal of radioactive materials; the management of radioactive patient, radiation monitoring dosimetry, and post-release radiation precautions for patients; as well as linac shielding and workload factors. The section was even involved in the Three Mile Island incident by evaluating the dosimeters that were used.

Radiation Safety in the New Millennium

In the new millennium, in parallel with the many new developments in using radiation in the diagnosis and treatment of cancer at Memorial Hospital, the Medical Health Physics section continues to be challenged in advancing radiation protection procedures. These developments include the implementation of PET/CT, the construction of the new Radiochemistry and Cyclotron Facility, and the use of novel radiolabeled monoclonal antibodies. To more accurately evaluate organ doses for specific diagnostic and therapeutic procedures, initiatives are underway using Monte Carlo methods and measurements in physical phantoms. These studies will likely provide data for future clinical epidemiology studies.

Institutionally, the Radiation Safety section developed inter-departmental collaborations. For example, recognizing the need of effective communication on the benefits and risks of procedures involving radiation, the Radiation Safety section is collaborating with the Departments of Radiology and Psychiatry & Behavioral Sciences to develop educational materials for patients, staff, and the general public. In a pilot epidemiological project, collaborating with the Department of Epidemiology and Biostatistics, we are studying a cohort of over 25,000 Memorial Hospital radiation workers with dosimetry data recorded over 70 years. This information will make an important contribution to the ongoing Million Person Study, a national NCRP initiative led by John Boice, Ph.D.

In the new century, medical health physicists at Memorial Hospital continue to lead in the field of radiation protection. Lawrence Rothenberg chaired the sub-committee (SC) for the NCRP Report No. 149 on mammography and other breast imaging procedures. That report, published in 2004, became the highest-selling publication in the history of NCRP. The NCRP Report No. 155, published in 2006, which provided the scientific basis for patient release from confinement with the appropriate precautions, was chaired by St. Germain, with Pat Zanzonico as a member. This report established the standard radiation safety procedures for managing patients with therapeutic administrations, and these procedures were implemented at Memorial Hospital by Dauer and Williamson. The 2008 NCRP Report No. 159 on the risk to the thyroid from ionizing radiation involved the efforts of Pat Zanzonico and Michael Tuttle, M.D., of Memorial Hospital's Endocrine Service. This report included information from results of the Chernobyl accident, with the important discovery that radiation, whether from external or internal sources, can increase the risk of thyroid cancer, with age at the time of exposure being the most influential factor, i.e., children are much more sensitive than adults.

In 2005, Dauer and Michael Zelefsky, M.D., from Memorial Hospital's Radiation Oncology Department participated in developing the radiation safety guideline for permanent implant of prostate cancer, resulting in the publication of ICRP Report 98. Dauer was subsequently elected to serve on the ICRP Committee 3 on Radiation Protection in Medicine for more than seven years (2010–2017). During that time he was the co-chair or member of several task groups. These included the ICRP Publication 139 on occupational radiological protection in interventional procedures, the Task Group 89 on occupational protection in brachytherapy procedures, and the Task Group 101 on radiological protection in therapy with radiopharmaceuticals. Within the NCRP, Dauer assisted on post 9/11 radiation safety issues, e.g., the 2011 NCRP Commentary No. 22 on the use of active monitoring systems for the detection of radioactive-threat-materials. In a collaborative effort of the Society for Interventional Radiology and the Cardiovascular/ Interventional Radiological Society of Europe, Dauer led the development of guidelines for occupational radiation protection of pregnant patients and workers in interventional radiology. He co-chaired the committee for the NCRP Commentary No. 26 (2016) that established a lower occupational dose limits for the lens of the eye. He also co-chaired the committee for the NCRP Commentary No. 27 (2018) evaluating the implications of recent radiation epidemiological studies on the linear-nonthreshold model. This important report is the basis of the new system of radiation protection in the United States. In 2014, Dauer was elected to the board of the NCRP and he continues to serve in that capacity, with special interest in low-dose radiation effects and dose reconstruction for former radiation workers and patients for epidemiologic studies.

Radiation Biophysics
Arrives at MSKCC

In the 1960s, Laughlin initiated research in radiation biophysics and radiobiology within the Sloan-Kettering Institute.

Early Research

Early collaborators with Laughlin included Louis Zeitz, Ph.D., who studied elemental analysis using soft x-ray absorption; Ira Pullman, Ph.D., involved in ESR study using nucleic acids and other biologicals; Harold Morrison, Ph.D., who assessed the role of SH-compounds in radiobiology; and Max Eidinoff, who focused on the mechanism of RSV (Rous Sarcoma Virus) replication and anti-metabolites.

In the mid-1960s, other scientists engaged in radiobiological research at ultra-high dose-rates at MSKCC were Edward R. Epp, Ph.D., Jae Ho Kim, M.D., Ph.D., plus colleagues. The research involved irradiation of thin layers of bacterial and mammalian cells at dose rates exceeding 10^{10} Gy per second. The radiation beam consisted of 600 keV electrons produced by a field-emission device and delivered in single pulses of 3–30 nanoseconds. The basic idea was to deliver the radiation in a short time with respect to the time-scale of radiochemical processes in cells occurring immediately after irradiation, eventually leading to cellular damage or repair of such damage. The study had particular relevance to the mechanism of the oxygen effect as it relates to cancer treatment.

In 1968, Epp, Herbert Weiss, and Ann Santamasso showed that such high-dose irradiation could generate a series of breaking survival curves for the bacterium E coli B/r when irradiated in a thin layer under various oxygen concentrations. This novel observation was attributed to oxygen depletion through radiation-induced intracellular chemical reactions by the first part of the electron pulse. Since oxygen replenishment by diffusion into the cell could not occur within the time duration of the 3–30 nanosecond pulse, the cells responded hypoxically to the dose delivered in the latter part of the pulse.

To confirm this explanation, E coli B/r was irradiated in a sealed vessel at conventional dose rates. In this experiment, breaking survival curves for E coli B/r were also obtained, as the cells were suspended in the "sealed" vessel such that the intracellular oxygen could not be replaced after being depleted by the radiochemical reactions. The "breakpoint dose" was determined to be linear with oxygen concentration.

These findings at ultra-high dose rates led to a double-pulse experiment. Two new field emission sources were used, each delivering a 3 nano-second pulse, cross-firing at a thin layer of bacterial cells. The first pulse was designed to deplete the intracellular oxygen, and the second pulse applied after a designed time delay, from microseconds to milliseconds. The amount of oxygen re-diffused to critical sites within the cell during the time delay was inferred from the known response of cells to single short pulses. In this way the measurement of the diffusion of oxygen into cells to various concentration levels was measured over six decades. It was found that significant oxygen concentration succeeded in reaching critical intracellular sites in about 10^{-4} s, which may be taken as an upper limit to the lifetime of oxygen-sensitive species suspected of being induced at critical sites in bacterial cells and perhaps directly involved in the basic mechanism underlying oxygen-induced radiosensitization.

The results were compared with a theoretical model developed by Kessaris, Weiss, and Epp in 1973, which yielded a value for the cellular oxygen diffusion coefficient of 0.4×10^{-5} cm^2 per second, a value one-fifth that of the generally accepted value for water. Similar results were obtained by applying the above techniques to the bacterium *Serratia marcescens*, thus confirming the generality of the effect for two bacterial strains.

Attention was next directed to the response of mammalian cells to single pulses of electrons at ultra-high dose rates. Similar breaking survival curves were observed when HeLa cells were thus irradiated, with the data again consistent with the physicochemical mechanism invoked above for bacterial cells involving the radiochemical depletion of oxygen. These data demonstrate the universality of the observed phenomenon.

Dosimetry for thin biological samples irradiated by nanosecond electron pulses at ultra-high dose rates presented nontrivial problems both in the determination of the absorbed dose per pulse and in the monitoring of pulse-to-pulse variation. Solutions to these problems were studied and published in a series of papers by McDonald, Weiss, Pinkerton, Epp, and Ling.

In 1963, Jae Ho Kim, M.D., Ph.D., had joined Dr. Max L. Eidinoff's laboratory in the Sloan-Kettering Institute's Biophysics Division. Eidinoff, a renowned physical chemist, had pioneered the development of tritium labeling of biologicals, including nucleic acid bases (e.g., thymine, uracil) in the early 1950s, pursuing the influence of anti-metabolites of nucleotides and nucleosides on the replication of Rous Sarcoma Virus using chick embryos. One of the first programs that Kim addressed was to determine whether the recovery and repair of sub-lethal radiation damage in mammalian cells would be inhibited by inhibitors of DNA or RNA synthesis using anti-metabolites.

Shortly after Eidinoff's retirement in 1965, Kim assembled a team of investigators (Drs. S. H. Kim, Alan Gelbard, Bozidar Djordjevic, and Amaury Perez) to study the role of metabolic and biochemical events occurring during the cell cycle of HeLa cells and how these events would play a role in the repair process of potentially lethal damage to irradiated synchronized cells. On the basis of the findings, the team identified novel metabolic inhibitors that would selectively enhance the radiosensitivity of tumor cells. Some of the inhibitors identified (e.g., chloroquine, autophagic inhibitor) have been of clinical interest and are in clinical trials in combination with chemotherapy or radiotherapy.

Other important collaborative studies were carried out by Dr. L. J. Old and his tumor immunobiology group. The radiobiology group played a key role in the identification of L-asparaginase-sensitive human neoplasms, a study that helped lead to a successful application of the E. coli enzyme in the treatment of acute leukemia.

In the mid-1970s, Drs. Kim and Eric Hahn initiated a series of *in vitro* and *in vivo* tumor radiobiology studies to determine the efficacy of combining hyperthermia with radiotherapy. On the basis of exciting preclinical data, the team conducted the first modern clinical trial combining mild heat and radiation in patients with superficial tumors (mycosis fungoides, malignant melanoma and breast cancer). At the same time, the team carried out a series of studies to determine the influence of the tumor microenvironment on the thermosensitivity of tumor cells— including pH, oxygen, and energy status—studies that led to the identification of novel hyperthermic sensitizers.

In the 1980s, Kim's group continued to develop non-hypoxic radiosensitizers that would be readily applicable in the clinic. On the basis of the cellular hyperthermia program, the team found that some of the hyperthermic sensitizers would equally enhance the radiation response of *in vivo* tumors (lonidamine, gossypol, fludarabine phosphate, thioguanine, etc). Phase I and II studies were carried out with MSKCC's Neuro-oncology and Developmental Therapeutics program.

Further Development in Radiation Biophysics

Upon his return to MSKCC in 1989, Ling moved his research laboratory from UCSF, relocating it to the 2nd floor of the Kettering Building, exactly where he began his career in 1971. He also transferred his research grants, one from NIH/NCI on "Radiosensitization of Mammalian Cells at Low Dose Rate" and one from the U.S. Department of Energy on "Activation of Oncogenes by Radon Progeny and x-rays." In the research for "Activation of Oncogenes by Radon Progeny and x-rays," Ling was happily reunited with Weiss. Together, they irradiated mammalian cells with alpha particles from the MSKCC cyclotron to simulate the effect of radon emission in causing lung cancer in humans. A talented physician-scientist from China, C-H Chen, M.D., Ph.D., joined them in this effort and reported the presence of a point-mutation in the N-ras gene in radiation-transformed rat embryo cells. Chen subsequently became Chair of Radiation Oncology at the University of Toledo, Ohio.

During that period, research at Ling's laboratory included studies on oncogene-induced radioresistance, radiosensitization by hypoxic cell sensitizer at low-dose-rate irradiation, and neoplastic transformation of oncogene-transfected rat embryo cells by gamma rays or 6 MeV alpha particles. Another research topic subsequently pursued was on radiation-induced apoptosis, its effect on cellular survival curves, and its relation to GAAD153 down-regulation.

In the last few years of the 20th century, Ling's laboratory gradually shifted its focus to biological imaging.

Another important contributor to the radiation biology program in medical physics was Gloria C. Li, Ph.D. During the 1970s at Stanford University, she carried out research on cancer treatment by pi-meson (pion), as well as research on hyperthermia cancer therapy. She designed and implemented a 3D treatment planning system for focusing the negatively charged pion particles from the 60-channel Stanford Medical Pion Generator into the tumor. Her hyperthermia work dealt mainly with a phenomenon termed thermotolerance, by which cells become resistant to higher temperatures, and thus acquire impunity to hyperthermia therapy. Her research led to the development of a model that accounts for the induction, development, and decay of thermotolerance in cells.

In 1981, she was recruited to the University of California, San Francisco. She continued her research on cellular responses to heat-shock and radiation, and rapidly rose to the rank of full professor. She pioneered functional studies of the mammalian heat-shock protein 70, and elucidated the mechanistic basis of thermotolerance. In 1990, she joined Memorial Sloan-Kettering Cancer Center (MSKCC) as Head of the Laboratory of Radiation and Hyperthermia Biology, was appointed Member with Tenure of Title at MSKCC, and concurrently served as Professor of Physiology and Biophysics at the Weill Cornell Medical College. There, while unraveling the mechanism of the heat-shock response in cells, she discovered the involvement of a dimeric protein Ku70/Ku80 in heat-shock protein regulation. The Ku protein dimer and a catalytic subunit termed DNA-PKcs are known to form the DNA-dependent protein kinase (DNA-PK), which binds to the ends at DNA double stranded breaks to activate their repair. Her laboratory generated knock-out mice deficient in the individual components of the DNA-PK complex, and these mutant mice were then used to elucidate the roles of the different components in DNA repair, VDJ recombination, tumor suppression, and radiosensitivity modification.

The modification of the radiation response by Ku70/Ku80 also led her to explore the possibility of radiosensitization of cells through a reduction of intracellular Ku70 protein concentration, or through overexpression of a fragment of Ku70 that competes with intact Ku70 subunit in DNA-PK for DNA end binding. Subsequently, she developed animal models containing quadruple reporter genes for studying hypoxia-induced gene expression, for non-invasive imaging of tumor hypoxia, and for hypoxia-targeted gene-radiotherapy. Using the animal models her laboratory had generated, *in vivo* hypoxia gene signatures were obtained for both acutely and chronically hypoxic cells, and these signatures were shown to be of prognostic value in the analysis of clinical trial results.

Significant research efforts and seminal findings on radiation biology conducted by Zvi Fuks and others in the Department of Radiation Oncology are not included in this book.

Computer-Assisted Treatment

Punched Cards and Tabulating Machines

Memorial's K. C. Tsien was considered to be the first person to use digital computers for calculating the radiation dose distribution for treatment planning in 1955. Instead of the conventional time-consuming process of reading off percentage depth doses from the overlapping isodose charts and adding them up for selected points, he used punched cards with sorting and tabulating machines.

The percentage depth doses produced by an x-ray field were first recorded on several sets of punched cards. The cards were then fed into the automatic tabulating machine, which summed up the data for all the points, tabulated the results, and plotted the findings.

In 1958, Richard Nelson and Mary Lou Meurk, also of Memorial Hospital, published a paper that used a similar approach for implant dosimetry.

Early Role for Computers

The role of computers in developing dose distribution plans for external beam therapy and interstitial and intracavitary therapy continued to be refined at Memorial. These newer methods produced dose distributions in two and three dimensions, accounting for surface curvature, internal inhomogeneities, and beam shaping devices. Such dose distributions were then used routinely in clinical practice at Memorial.

Memorial's Time-Sharing Computation Service

Due to the high cost of computers and software development, plus the required support, only a few large centers could afford computer-aided treatment-planning tools. In 1967, led by Garrett Holt, Memorial established the Memorial Dose Distribution Computation Service (MDDCS). With it, a participating hospital, using an ordinary teletype as a remote terminal, could access a time-sharing computer service (operated by ITT) via a telephone line through a modem to input patient data and beam configurations, perform dose distribution computations using Memorial software, and retrieve the resulting treatment plan. Dose distributions were printed on the teletype in the form of dose matrices and character maps depicting dose distribution. Initially, five institutions participated in the Memorial network.

In 1971, Radhe Mohan, Ph.D., joined Memorial's Department of Physics. He was charged with refining and upgrading external beam and brachytherapy treatment planning systems under the guidance of Garrett Holt, Alan Baker, Steven Balter, and Lowell Anderson. In addition to improving the dose computation algorithms, he incorporated newly available devices, including faster terminals and modems, digitizers to facilitate entry of patient contours, and digital plotters to produce graphical displays of dose distributions.

In 1975, the physics department decided to establish an in-house computer network, replacing the time-sharing service. Initially, the new system was based on a dual (for redundancy) DEC PDP 11/45, which was later upgraded to a PDP 11/55 and, starting in the early 1980s, to a DEC Vax series of computers. Under Mohan's leadership, the network grew rapidly, peaking at over 100 simultaneous user institutions. At one time or another over its 25-year life span, the network served the external beam and brachytherapy treatment planning needs of more than 200 institutions throughout the United States.

With the emergence of new, affordable commercial treatment planning systems using mini- and micro-computers with advanced graphic capabilities, the need for the dose computation service began to diminish, with usage declining in the latter part of the 1980s. The service was discontinued by the mid-1990s.

New Treatment Planning and Data Management Systems

Starting in the late 1970s, in addition to the continued effort to enhance treatment planning capabilities, Mohan and his colleagues developed the first system for computer-aided dose-distribution measurements which allowed the acquisition of the large amounts of data necessary for commissioning treatment planning systems for accurate dose computations. This system was also used by various MDDCS participants to assist in acquiring necessary data.

Also, this same period saw the development of a CT-based 2D external beam planning system based on the existing external beam dose computation system. Although the system saw limited clinical use, it became the foundation for the 3D CT-based external beam planning system developed in the 1980s. Mohan and colleagues also developed the first system for recording and verifying radiation treatment setup parameters to assure the correctness and safety of treatments. The system continued to evolve and became a part of a larger patient data management system. For almost 20 years, it was the only system that could be interfaced with treatment machines of multiple vendors. Eventually, commercial vendors adopted the concepts behind both the dosimeter system and the record-and-verify system.

Radiation Dose Calculation

A significant fraction of the computer applications in Memorial's effort during the 1980s was devoted to research based on Monte Carlo techniques to simulate radiation transport. The goal was to improve understanding of the characteristics of radiation beams emerging from the treatment machines, their interactions with beam-shaping devices, and the mechanisms underlying dose deposition in complex heterogeneous patient anatomy.

The results led to the continued evolution and improvement of the accuracy of computed dose distributions over the next two decades at Memorial and elsewhere. Based on this knowledge, three-dimensional dose distribution computation methods, a 2D pencil beam Fast-Fourier Transform-based convolution method, and a Differential Pencil Beam method were developed.

A variation of the latter, termed the "superposition-convolution" method, developed by Rock Mackie and his team, was used in a commercial treatment planning system. The 3-D Mackie method was added to Memorial 3D and IMRT treatment planning systems and was in use until 2012. Starting in the early eighties, the PDP 11 computers were replaced with DEC Vax 11/780 computers, which were gradually upgraded to more powerful systems as computational needs grew. Beginning in the 1990s, the Vax series of computers were replaced with DEC Alpha series.

Evolution in Computer Platforms

By the mid-1990s, computing philosophy gradually changed from large, centralized systems to small, distributed systems, according to Chen-Shou Chui, Ph.D., who succeeded Radhe Mohan as Service Chief. In keeping up with the trend, the team accordingly switched from the large VAX/VMS systems with Lexidata displays to AlphaStation/X-window systems to take advantage of the computing power and the graphics display capability of the latter. This allowed each planner to have his/her own dedicated workstation to design treatment plans more efficiently.

3D Conformal Radiation Treatment Planning

Going back a few years to 1984, through the efforts of John Laughlin, Memorial along with three other institutions—Mass. General Hospital (MGH), Washington University, St. Louis, and the University of Pennsylvania—were awarded contracts to "evaluate" their respective 3D conformal radiation treatment (3DCRT) planning systems. Unfortunately, with the exception of MGH, which had a version developed for proton therapy, none of others had 3DCRT planning systems.

Nevertheless, with tremendous effort on the part of the computational physics team, along with the resources and support provided by the leaders of the departments of Medical Physics and Radiation Oncology, a working version was developed within a year of the contract award. It used sophisticated graphic displays for real-time visualization of patient anatomy in 3D, dose distributions overlaid on CT images, and array processors and numerous algorithms to accelerate computation of advanced 3D dose distributions.

Within three years of the contract, in early 1988, the Memorial system emerged as the leading, most comprehensive 3D conformal radiation treatment (3DCRT) planning system implemented for clinical use. In addition to its use in routine practice, it has been used in a number of important clinical trials due mainly to the persistent efforts of Ling and Zvi Fuks.

Multi-Leaf Collimators

Another development that coincided with the evolution of 3DCRT planning systems was the development of multi-leaf collimators (MLCs) by Varian, in part at Memorial's urging. The shaping of static fields by MLCs and the determination of leaf positions were incorporated into the treatment planning systems. These positions were then transferred to the treatment machine computers for delivery.

Led by Ling, the Department of Medical Physics applied for and was awarded an NCI P01 (Program Project Grant) in 1991. The P01 research was focused on the ongoing 3DCRT research and development effort and its clinical applications. Early work related to computers in radiotherapy during the P01 period included research on novel methods of optimizing radiation treatment plans using techniques such as simulated annealing and employing objective functions-based dose-volume-histogram, Tumor Control Probability, and Normal Tissue Complication Probability. These developments laid the foundation for implementing intensity-modulated radiation therapy (IMRT).

In 1993, during a visit to Memorial, Thomas Bortfeld, Ph.D., of the German Cancer Center in Heidelberg (DKFZ), collaborated with Mohan to integrate an IMRT optimization (inverse planning) algorithm he had developed into the Memorial 3DCRT planning system. Results were astonishing, and it was the beginning of the IMRT era.

This was followed by the development of algorithms by Chen Chui and colleagues to transform the optimized nonuniform intensity distributions into trajectories of leaves of dynamic MLC. Simultaneously, Ling and Fuks prevailed upon Varian's leadership to produce the dynamic MLC control system by which the leaves move to predefined positions as a function of dose.

Computer-Controlled Treatment Delivery

A further development in the early 1990s was the introduction of so-called "segmental therapy," in which MLC-shaped radiation fields, called segments, were set up and treated under computer control. The degree of operator intervention varied from one in which a setup of each segment is initiated by the operator, to one in which a sequence of segments is delivered without therapist intervention.

Computer control consisted of setup not only for radiation delivery, but also for mechanical components, including gantry and collimator angles, collimator jaws, MLC, and treatment couch position. At that time, the MM50 medical microtron by the Swedish firm Scanditronix Medical was the only commercially available treatment machine with this capability.

Treatment with the microtron at Memorial used two computers: a control computer proprietary to the treatment machine manufacturer and an external computer running software developed by Mageras, Mohan, and coworkers for this purpose. The external computer transferred instructions from the treatment planning system to the microtron, verified the correct setup of the machine, and recorded the treatment. Clinical use of segmental therapy focused on prostate conformal treatments, in which six conformally shaped fields were set up and treated within four minutes.

Clinical Implementation of IMRT

During the same period, research focused on the algorithm development and clinical implementation of IMRT, including the following components:

Optimization. In the early 1990s, Bortfeld brought his optimization program to MSKCC, which was then implemented by Mohan and Xiao-Hong Wang as a stand-alone program. Bortfeld's optimization algorithm was dose-based. In the meantime, a graduate student from Columbia University, Spiridon Spirou, was working on his Ph.D. thesis under the supervision of Chen Chui. Spirou's thesis work involved the optimization and delivery of IMRT. Spirou's optimization algorithm was different from Bortfeld's method in that it was dose/volume-based and especially suitable for certain organs such as lungs and liver.

Delivery. In addition to the optimization algorithm, Spirou and Chui also worked on the delivery of IMRT using conventional MLCs, then available on Varian's machines at MSKCC. The delivery methods worked out by Spirou and Chui included both dynamic mode (DMLC) and segmental mode (SMLC or step-and-shoot). The final leaf motions were written to a computer file (so-called 'DVA' file) and sent to the Varian machines for execution.

Dose calculation. For the nonuniform intensity distribution determined by the optimization algorithm and delivered by the MLC, a dose calculation method was needed to determine the resulting dose distributions in patients. Chui modified the then existing pencil-beam model on the planning system for IMRT dose calculations. The pencil-beam model was fast and reasonably accurate.

Independent secondary check for IMRT. For the clinical implementation of IMRT, Chui also developed an independent program that reconstructed the intensity distributions from the 'DVA' files, including a separate dose calculation algorithm. This calculated result was compared with the original result for verification.

Dosimetry verification. Extensive measurements were made by Tom LoSasso of the Dosimetry Section and Chui to confirm the accuracy of the entire IMRT process: comprising optimization, delivery, and dose-calculation. Dosimeters used were primarily ion chambers and film. In general, in a homogeneous phantom, accuracy within ±2% was achieved.

On September 17, 1995, all these efforts culminated in the world's first IMRT delivery with a conventional MLC on a Varian 2100C machine. Shortly after, the first IMRT patient received a 9-Gy boost phase treatment for prostate cancer. The intensity distribution was obtained by Bortfeld's algorithm and the delivery was achieved with DMLC.

Around 1997, after extensive testing by the treatment planning group, Spirou's dose/volume-based optimization algorithm was integrated into the planning system, then available for routine clinical use. Initially, IMRT treatments were primarily used for prostate patients, but quickly thereafter were applied to other sites.

In the late 1990s, simplified IMRT methods for breast were developed and implemented, greatly improving the efficiency and quality of intact breast treatment.

By the end of the decade and in the early 2000s, the PC industry had made great progress and gained widespread popularity. The computing environment was again changed from the Alpha Station/X-window system to the PC-window system.

Electronic Image-Based Treatment Guidance and Verification

The early and mid-1990s saw an increased interest in electronic portal imaging devices (EPIDs) for verifying patient positioning and radiation field shaping at treatment. Potential advantages of EPIDs over film included the ability to rapidly acquire and view the images and to apply computerized techniques to enhance the images and extract quantitative information from them.

An essential ingredient for their successful use in the clinic, though lacking at that time, was a picture archiving and communication system (PACS) specifically designed for radiotherapy departments. In the mid-1990s, the computer service group developed and implemented such a system at Memorial, which facilitated the increased clinical use of EPIDs and the eventual replacement of film with electronic images.

The advent of treatment-room kilovoltage imaging systems in the early 2000s spawned interest in image-guided radiation therapy (IGRT) in the radiation oncology community. In response to this new need, the computer service group developed image registration algorithms for use with cone-beam CT, which allowed three-dimensional visualization of the tumor in the treatment room and correction of patient position to improve radiation targeting accuracy. The software included a rigid-body registration algorithm, which was implemented for routine clinical use, and a deformable registration algorithm which, due to its complexity, was used for research only. In addition, compatibility with DICOM was developed, allowing images and other treatment data to be transferred between different systems.

Medical Physics Computing Service for Radiology and Radiation Oncology

As presented in another section, Peter Kijewski was recruited in 1994 to explore PACS development for radiology. Subsequently, support provided to radiology in computing technology was expanded to include two other faculty members, B. Zhao, Ph.D., and Luc Bidaut, Ph.D.

Around 2002, recognizing the potential synergy between the two groups of computer scientists, both homed within Medical Physics, but separately supporting the Departments of Radiation Oncology and Radiology, the groups were combined under the aegis of Medical Physics Computer Service with Chen Chui as Service Chief. Upon Chui's departure, Gig Mageras served as Service Chief. This structure lasted until the early 2010s.

Volumetric Modulated Arc Therapy

Around 2007, attention in the radiation oncology community focused on a new type of IMRT delivery, called volumetric modulated arc therapy, or VMAT. In contrast to IMRT, which involves motion of MLC alone during dose delivery, VMAT involves the simultaneous motion of MLC, gantry rotation, and variation in radiation dose rate.

The principal advantage of VMAT over IMRT is the decreased treatment time, thus reducing the likelihood of patient motion and loss of targeting accuracy during treatment. Because of the additional treatment machine motions involved, VMAT posed new challenges for treatment plan optimization and dose calculation. Perry Zhang and coworkers in the Computer Service developed new algorithms for VMAT treatment planning, which were incorporated into the in-house treatment planning system. This capability was used for a few years at Memorial for treatment of tumors in prostate, lung, spine, and pancreas, until the acquisition of a commercial treatment planning system with similar capabilities.

Computer-Assisted Brachytherapy

In the 1970s and 1980s, Mohan and Anderson extensively applied computer technology to brachytherapy planning and dosimetry. The effort included the incorporation of shields for cervical applicators into dosimetry calculation, and development of nomograms.

In the late 1990s, Marco Zaider, Ph.D., and Michael Zelefsky, M.D., in collaboration with the Computer Service developed an intraoperative computer-based three-dimensional conformal optimization capability for ultrasound-guided radioactive seed implants in prostate. The principal advantage of this methodology over preoperative CT-based implant planning was the ability to further enhance the dose distribution of the target and minimize dose to normal tissue structures, confirmed in post-implant analyses of patients.

In the mid-2000s, Zaider and Zelefsky developed a methodology for treatment planning that combined low-dose-rate radioactive seed implants in the prostate with high-dose-rate IMRT. The rationale for this approach was that by computing a fused dose distribution, the synergism between the biologic effects of the two modalities could be better realized than by designing treatment plans independently of one another. Implementation of the treatment planning algorithms was carried out by Chui and coworkers in the Computer Service. A study of over 100 patients showed the combined modality treatment to be associated with low rates of normal tissue toxicities.

Subsequently, the introduction of a mobile cone-beam CT unit in the operating room has enabled the acquisition of three-dimensional images of the patient for prostate implant evaluation. An important component for utilizing this capability was the development of software tools to automatically identify the implanted seeds in the cone-beam CT images, transfer of the seed locations to a post-implant ultrasound image, and calculation of the dose to target and normal tissues, thus permitting evaluation within minutes.

Nuclear Imaging Physics

In the late 1960s, the field of what was to become Nuclear Medicine was about to enter a period of dramatic development and growth. The Chairman of Medical physics, John S. Laughlin, was at the same time "chief" of the Biophysics Division of SKI and of the Radioactive Isotope Service within Memorial Hospital. The Biophysics programs included the development and production of novel radioactive tracers, especially those utilizing the biologically important elements carbon, nitrogen, and oxygen. The short half-lives of these radioisotopes necessitated a dedicated in-house cyclotron for their production. Such a cyclotron also permitted studies with other nuclides that would otherwise not have been available. Concurrently, instrumentation was being developed to image and quantify these tracers.

A rectilinear scanner, the High Energy Gamma (HEG) scanner, with large detector crystals and extra thick shielding to provide adequate sensitivity and shielding for the high-energy gamma rays and annihilation photons, was designed and built in-house for patient research. Additionally, technology was developed to record data digitally and process these on an IBM1800 computer rather than to just store an image on film, as was then the norm. A second version of this with somewhat less shielding, the Multi-Function Scanner (MFS), was built for routine use in the clinic, which permitted scanning with the patient in various positions—lying, sitting, or standing.

A single-head, small field-of-view version of the gamma camera was just becoming available and would be acquired for the clinic. In the meantime, it was decided to build a slightly larger field camera with a thicker, more sensitive scintillator and heavy shielding to image the 511 keV photon from positron-emitting nuclides. This camera would have two opposed heads and could be operated either in singles mode with high-energy collimators or without collimators in coincidence mode, allowing longitudinal tomography. To permit dynamic imaging, this device, Total Organ Kinetic

73

Imaging Monitor (TOKIM), was connected directly to the IBM 1800 computer.

A suite of computer programs was developed that could accept whole-body data or small-field static or dynamic data from any of the research or clinical devices and process them. Output was in printed form using symbols that provided both a visual representation of the scan and provided quantitative information.

Meanwhile, the Radioactive Isotope Service provided the routine clinical services under the auspices of a medical physicist with three physician representatives from the Departments of Radiology, Radiotherapy, and Medicine. The imaging equipment consisted of the Multi-Function Scanner, two commercial whole-body scanners (with MSK digital acquisition systems), and a small-field static gamma camera. There was also a thyroid laboratory with an uptake probe and sample-counting equipment. Radionuclides were mostly purchased commercially, with one exception. Fluorine-18 ($Na^{18}F$) was produced on the in-house cyclotron for bone scanning using the whole-body rectilinear scanners. This two-hour half-life radionuclide was also shipped to outside hospitals as far away as Philadelphia.

The clinical part of the Radioactive Isotope Service was renamed Nuclear Medicine Service and transferred to the Department of Medicine under Richard Benua, M.D., around 1970. The part of the Radioactive Isotope Service remaining in Medical Physics consisted of nuclear medicine physicists and physics operations, plus the Radioactive Isotope Laboratory. The Biophysics section (imaging and cyclotron) remained as before.

The Nuclear Medicine Service was transferred to Department of Radiology under Steven Larson, M.D., in 1978. The Biophysics section was also placed under Dr. Larson.

Radionuclide Therapies

Radioiodine Therapy

In the 1940s, the importance of ^{131}I for the treatment of benign and malignant thyroid diseases became realized. Knowledge of how much radioiodine to administer to adequately treat the disease was sought. In the 1940 and 50s, Leonidas Marinelli, an Argentinian-born physicist came to MSKCC, where he derived the first thyroid dose formula, that sometimes is still used today. His landmark papers—"Dosage determination with radioactive isotopes" with Quimby and Hine in *Nucleonics* in 1948 and "Radiation dosimetry in the treatment of functional thyroid carcinoma with I^{131}" with Hill published in *Radiology* in 1950—are classic papers that were to serve as a foundation stone of radionuclide dosimetry and the later Medical

Internal Radionuclide Dosimetry (MIRD) committee developed by Lovinger and Berman.

Using the Marinelli formula, one can calculate the activity of ^{131}I needed to achieve the desired radiation absorbed dose from a measurement of the peak radioiodine uptake in the thyroid, and an assumed thyroid disease mass and effective half-life in the thyroid tissue.

Measurement of the rate of elimination of ^{131}I from the body via the urine was a common procedure at MSKCC. Marinelli invented a special beaker to measure radioactive fluids. It consisted of a regular laboratory beaker with a cylindrical hollow tube at the beaker center. A radiation detector was inserted in the hollow tube to measure the radioactivity in the urine or any other radioactive fluid contained in the beaker.

Further refinements to thyroid dosimetry were made by Benua and colleagues, who added measurements of the pharmacokinetics of radioiodine in the blood using a high-sensitivity well scintillation counter, as well as measurement of the elimination of ^{131}I by whole-body counting. The collection of this data and its use in the determination of the maximum tolerated dose of radioiodine to minimize bone marrow toxicity became known as the Memorial method of thyroid dosimetry. This approach has been employed at MSKCC for the radioiodine management of metastatic thyroid cancer for over 60 years, without a single serious instance of marrow toxicity.

Radiolabeled Antibodies

The appointment of Steve Larson, M.D., as chief of Nuclear Medicine in 1988 brought a new focus on targeted therapies beyond the treatment of thyroid cancer. Dr. Larson has a long-standing interest in the development of radiolabeled antibody therapy. MSKCC was the right environment in which to have an outstanding Department of Immunology under the leadership of Herb Oettgen, followed by Alan Houghton, with sustained funding support by an NCI program project grant. In addition, the brilliant physician/scientist Lloyd Old was the William E. Snee Chair of Cancer Immunology at MSKCC from 1971 to 2011, and he served as the founding scientific and medical director of the Ludwig Institute for Cancer Research (LICR), with a dedicated focus on cancer immunology. Antibodies developed by these programs could be radiolabeled with high-energy β-emitters such as ^{131}I or ^{90}Y and then injected into cancer patients. The promise of radioimmunotherapy was the ability to target disease not accessible by other methods, and to treat metastatic and micrometastatic disease.

With the development of radioimmunotherapy emerged the need for improved radionuclide dosimetry. George Sgouros, Ph.D., who began as a volunteer high school student at MSKCC, returned in 1989 as a post-doctoral trainee to adapt external beam radiotherapy planning methods to radionuclide therapies. In 1990, Sgouros published a landmark paper in *Nuclear Medicine* entitled "Treatment planning for internal radionuclide therapy: three-dimensional dosimetry for non-uniformly distributed radionuclides," which described an approach to calculate the spatially varying radiation absorbed dose superimposed on CT images. This led to an advanced software tool called 3D-ID that allowed region-of-interest (ROI) drawing over critical organs on serial gamma camera images (SPECT or PET), and the generation of average dose and dose volume histograms to these organs. Another major advance was the derivation of an equation to calculate the red marrow-to-blood activity concentration ratio, based on the hematocrit and the red marrow extracellular fluid fraction of a patient.

During the late 1980s and 1990s, several new antibodies were developed at MSKCC: M195 from the Leukemia Service (Dave Scheinberg), 3F8 from the Department of Pediatrics (Nai-Kong Cheung), and A33 and G250 from the New York branch of the Ludwig Center of Immunology. All these were radiolabeled with either radioiodine or radiometals (e.g., ^{111}In) to allow imaging and to determine pharmacokinetics, biodistribution, and dosimetry. The Nuclear Imaging physicists, which included Marty Graham, Keith Pentlow, Farhad Daghighian, and George Sgouros, were actively engaged in these studies.

The Department of Energy (DOE) was a strong supporter of radionuclide production for medical applications. In the 1980s, John Laughlin obtained a substantial grant for the development of new radionuclides with potential for cancer imaging. The grant was passed on to Steve Larson upon his arrival at MSKCC. Ron Finn, Ph.D., was then recruited to manage the CS-15 cyclotron and the Radiochemistry Service. Ron pioneered the development of targets on the CS-15 for the production of an unusually long-lived ($\tau_{1/2}$ = 4.2 days) positron-emitting radionuclide, ^{124}I, that allowed the first PET imaging to be performed with antibodies. Keith Pentlow and Marty Graham proposed the use of quantitative imaging of ^{124}I using PET with applications to radioimmunodiagnosis and radioimmunotherapy in a seminal paper published in 1991. It was not long before the first clinical study was performed in a pediatric patient using ^{124}I-labeled 3F8 antibody by Nai-Kong Cheung and Steve Larson. ^{124}I and several other positron emitters did not undergo simple positron emission, but oftentimes emitted concomitant prompt gamma rays that could not be separated by the coincidence timing window. Keith Pentlow and Brad Beattie worked on the first successful correction schema to eliminate the artifacts resulting from the prompt γ-emissions accompanying positron decay that accompany several radionuclides, such as ^{124}I, ^{86}Y, and ^{76}Br. The increased accessibility of ^{124}I allowed more radioimmunoPET studies to be performed, most noticeably ^{124}I-A33 and ^{124}I-G250, led by Chaitan Divgi and Steve Larson in Nuclear Medicine and Joe O'Donoghue and John Humm from Medical Physics. These were the first studies to show that late (6–8 day post-injection) imaging by PET imaging resulted in accurate estimation of the antigen density within tissue as validated by quantitative digital autoradiography and antigen immunohistochemistry of the surgical resected specimens.

Another remarkable achievement of this period was driven by David Scheinberg from the leukemia service, George Sgouros from Medical Physics, and Mike McDevitt and Ron Finn from radiochemistry, who conducted the first phase I study of an antibody (M195) targeting leukemia with an alpha-emitter, ^{213}Bi. Since ^{213}Bi only possessed a 46-min half-life, it could only be used with blood-borne cancers. The first paper on the clinical use of a targeted α-emitter was published by Sgouros et al. in 1999 entitled "Pharmacokinetics and dosimetry of an alpha-particle emitter labeled antibody: ^{213}Bi-HuM195 (anti-CD33) in patients with leukemia." Continuation of the Immunology Program Project grant under Scheinberg's leadership, with significant medical physics input, led to the introduction of the concept of the α-particle-emitting radionuclide ^{225}Ac. This became known as the α-particle nanogenerator, because ^{225}Ac with its 10-day half-life decayed through 4 α-emissions until it reached its final stable lead daughter atom. ^{225}Ac labeled M195 was first used in patients in the early part of the 21st century. It continues to be under active investigation as a potential for targeted nanoparticle therapies.

Antibodies and nanoparticles are very high-molecular-weight constructs. At the other end of the molecular weight spectrum are the elements by themselves. ^{223}Ra administered as a simple salt, radium chloride, ^{223}Ra-Cl_2, was developed as a therapeutic agent by the Norwegian-based company, Algeta AS, to eliminate cancer metastases in bone. The first trial in the United States with ^{223}RaCl$_2$—known at the time as Alpharadin (today as Xofigo)—was conducted by Jorge Carrasquillo from Nuclear Medicine, Mike Morris from Medicine, and Joe O'Donoghue and John Humm from Medical Physics. This agent was subsequently shown in a multi-center trial to be therapeutically beneficial in castrate-resistant prostate cancer metastatic to bone.

The Growth of Nuclear Medicine

In 1993, John Humm came to MSK, initially in the dosimetry section headed by Chen Chui. In 1995, he was appointed section head of Nuclear Imaging Physics by Clif Ling. John Humm had a strong background in radiolabeled antibody dosimetry. He formed a powerful team of nuclear imaging physicists to support the strong clinical radioimmunotherapy program at MSK. In 1995, MSKCC had four Nuclear Medicine physicians: Steve Larson, Sam Yeh, Chaitan Divgi, and Homer Macapinlac, and four medical physicists: John Humm, George Sgouros, Farhad Daghighian, and Hovanes Kalaigian. Together they contributed to a period of unprecedented growth in nuclear medicine, such that by 2010 there were approximately 12 full-time nuclear medicine physicians and 10 medical physics personnel. The principal cause of this growth and influence was due to the introduction of whole-body positron emission tomography (PET).

Positron Emission Tomography (PET)

In October 1995, the first whole-body PET scanner (the GE Advance) was installed on the 2nd floor of the Schwartz building in the Laurent and Alberta Gerschel PET Center, with the generous support of the Gershel family. Initially, PET studies were often not reimbursed prior to demonstration that it was a reliable and valuable diagnostic tool. The bulwark of PET examination used fluorodeoxyglucose (FDG), as it still does today. Whereas there exist false positives associated with inflammation or brown fat, the immense benefit of detecting metastasis established PET as an important tool in cancer management.

In 1998, the first disease site to receive reimbursement for FDG PET was non-small-cell lung carcinoma due its significantly higher sensitivity and specificity for lesion detection relative to CT. Medicare reimbursement was soon expanded in 1999 to the diagnosis, staging, and restaging of esophageal, colorectal, melanoma, lymphoma, and head and neck (except CNS and thyroid) cancers. The GE Advance whole-body PET scanner soon became one of the busiest scanners in the country, performing about 12 PET scans per day. Yusuf Erdi, a post-doctoral trainee of John Humm, became the first dedicated PET medical physicist responsible for the technical aspect of the PET, with the support of a GE engineer, Jean Ghailly. In addition to supporting the GE Advance, Erdi worked on two challenging research problems. The first was how to determine the extent of a malignant lesion based on the average standard uptake value (SUV_{av}). This is of relevance for radiotherapy in that PET images are used in defining the target volume for radiotherapy planning. The second and related problem was how to define a tumor in the thorax in the presence of respiratory motion. The long (4 to 6 min) acquisition times of PET scans per 15 cm field of view did not allow for breath hold. Adopting the method developed by radiation oncology physicists, Erdi and a post-doc, Sadek Nehmeh, gated the PET scan acquisition, first demonstrating the method on an oscillating phantom and then applying it to clinical PET studies. This work led to significantly improved quantification of PET scans in the lung.

In 2001, Dr. David Townsend and his team from the University of Pittsburgh developed the first PET-CT scanner. Once again, skeptics demurred as to the waste of CT scanner time attached to a PET scanner. Yet, the technology rapidly gained acceptance. The ability to visualize a PET image overlaid on an anatomical map proved invaluable. Between 2001 and 2002, MSKCC bought two PET-CT scanners, the CTI/Siemens Biograph and the GE Discovery LS. The Biograph was initially delivered with a two-slice (duo) Emotion CT, soon to be upgraded to a four-slice CT scanner. The GE Discovery LS was equipped with a four-slice Brightview CT scanner. The Biograph used lutetium oxyorthosilicate (LSO) detector crystals, whereas the GE unit contained bismuth germanate (BGO). These scanners became so valuable in presurgical diagnosis and for monitoring treatment response that physicians no longer want to see PET emission scans without a concomitant CT.

However, the new multi-modality scanners did bring some new problems, the biggest of which was the mismatch between the PET and CT images of the lung. This was a consequence of the very different acquisition durations of the CT (10 to 20 seconds) and the PET acquisition (3 to 5 minutes, representing a superposition of multiple breathing cycles). Erdi and Nehmeh already had the solution. Acquire the PET as a respiratory gated study. At this same time, Gig Mageras coordinated a visit to MSK by Tinsu Pan from General Electric to demonstrate his new 4D CT protocols, first tested at MGH in Boston. With this protocol, the CT data was acquired for each multi-slice panel during one full respiratory cycle, allowing the changing lung anatomy and the motion of lesions within that anatomy to be captured. Synchronization of the respiratory phases from the 4D CT with the 4D PET allowed respiratory motion of both image sets to be observed in cine mode. The first 4D PET/CT acquisition was performed at MSKCC in 2003 by Nehmeh, Erdi, and Mageras, as well as Tinsu Pan from General Electric.

Molecular Imaging

At the new millennium, there emerged a new buzz word "molecular imaging," which evoked a new excitement about the potential of developing non-invasive imaging techniques to study cancer biology. The modalities leading the charge were nuclear medicine, magnetic resonance, and optical imaging. The field was fueled by the greater availability of PET and MR scanners, the emergence of small animal high-resolution imaging devices, and advances in molecular biology, in particular the development of reporter gene imaging. Ron Blasberg and Juri Gelovani (*aka* Juri Tjuvajev) at MSK were key innovators of this field, alongside of Sam Gambhir and Simon Cherry at UCLA. Clif Ling recognized the opportunity for Medical Physics to play a large role in the development of this field. His strong interest and background in radiobiology led him to form a leadership team to brainstorm on how to use molecular imaging for the purpose of defining radioresistant targets within the tumor for dose painting. The Medical Physics team consisted of John Humm, Jason Koutcher, Gloria Li, Joe O'Donoghue, and Pat Zanzonico. The radiation oncologist, Nancy Lee, and colorectal surgeon, Jose Guillem, were also key players. A very important concept paper entitled "Towards Multidimensional Radiotherapy (MDCRT): Biological Imaging and Biological Conformality" was published by Ling, Humm, Larson, Amols, Fuks, Leibel, and Koutcher in 2000, which defined the research direction. These ideas resulted in a successful period of grant funding, culminating in a program project grant on the topic of tumor hypoxia imaging. The research focused on the pre-clinical evaluation of hypoxia radiotracers and MR surrogates thereof, with immunohistochemical and partial oxygen probe corroboration of the different imaging approaches. This work successfully translated into the use of DCE MRI (perfusion and permeability) combined with fluoromisonidazole (^{18}F-FMISO) PET (hypoxia) in head and neck cancers (Lee, Dave, and Humm). In the 2010s, Lee and Humm pursued the concept of using hypoxia PET images in chemoradiation therapy management of patients, proposing dose de-escalation for patients with good prognosis HPV+ H&N cancers with the intent to reduce normal tissue complications.

In parallel to the advances in molecular imaging equipment for clinical use were commercial developments in small animal imaging equipment. Such equipment was already in place in the case of MRI. Small animal PET scanners were, however, entirely new. Simon Cherry developed the first, so-called "microPET" scanner at UCLA. The commercial version of this scanner was developed and marketed by Concorde Microsystems. MSK followed UCLA and Washington University in purchasing the third microPET scanner in 2002. As reporter gene technology grew, so did the imaging equipment to support it, most notably bioluminescent and fluorescent imaging systems. Pat Zanzonico, who was recruited from Weill Cornell in 1998, became one of the administrative heads alongside Jason Koutcher and Steve Larson. Pat is a pivotal leader of the small animal imaging core at MSKCC, which has since grown to become one of the largest in the world, containing micro-PET/CT, SPECT/CT, ultrasound, an array of optical imaging devices and, most recently, a microCT/orthovoltage radiotherapy unit for 3D conformal treatment planning. This core facility was consolidated in 2006 behind the animal housing barrier facility in the sub-basement of the new Zuckerman building. This core has proven to be financially robust on account of not only the wide spectrum of avid users from a large number of the SKI laboratory programs, but the fine administrative leadership of Pat Zanzonico.

C. Clifton Ling:
A New Era for Medical Physics

Early Years

Clifton Ling, Ph.D., was born during World War II in Guilin, Guangxi, China, in June 1942. In his tumultuous early life, with the Sino-Japanese War and the Chinese civil war, Ling's parents moved to Chungking, Shanghai and then in 1949 to Hong Kong. After his primary and secondary education, Ling boarded the SS President Wilson in 1962, crossed the Pacific Ocean, and landed in San Francisco.

Ling attended Oregon State University, graduating in 1965, and then the University of Washington, receiving his Ph.D. in Nuclear Physics in 1971.

C. Clifton Ling

MSKCC: Part One

Ling worked at MSKCC for two periods, 1971 to 1974 and 1989 until becoming Member Emeritus in 2012.

In looking for work after his Ph.D., Ling happened upon job advertisements for medical physics and radiation research. Ling then joined Edward Epp's Laboratory of Physical Biology at MSKCC in 1971. There Ling learned radiation physics and biology, attending courses given by Lowell Anderson, Ph.D., and J. H. Kim, M.D., Ph.D.

The Intervening Years

In 1974, Ed Epp was recruited by Herman Suit, M.D., to become Chief of Radiation Physics at Mass General Hospital (MGH), and he brought Ling to Boston as an assistant radiation physicist. In 1979, Ling moved to George

Washington University Hospital as an associate professor. In 1985, Ling joined UCSF as professor and Vice-Chair of the Department of Radiation Oncology.

MSKCC Part 2: Recruitment

In 1987, MSKCC began a search for Laughlin's successor. Around that time, Steve Leibel, M.D., who knew Ling at UCSF, joined MSKCC as Vice-Chair of Radiation Oncology, and sometime afterward called Ling requesting an updated CV. Subsequently, Zvi Fuks, M.D., called Theodore Phillips, M.D. (Chair of Radiation Oncology at USCF) to ask whether MSKCC might invite Ling to visit concerning the physics chair. Ling asked Phillips for his opinion, and Phillips's reply was, "You can't turn down a job which you have not been offered."

"The best job in the field of medical physics"

Ling visited MSKCC in June 1988 and met with Samuel Hellman, Zvi Fuks, and others on the search committee, including Lowell Anderson, who represented Medical Physics. Hellman was about to leave Memorial Sloan-Kettering for the University of Chicago, but he strongly urged Ling to go to New York as the position, in his words, "is the best job in the field of medical physics."

Ling was back for a second and third visit to Memorial, with interviews with Paul Marks, M.D., President of MSKCC, and meetings with the administration.

The Ling Years: Forward March

In November 1989, Marks formally offered Ling the position of Chairman. After further negotiation, Ling accepted the offer on January 14, 1989, and became Enid A. Haupt Professor and Chairman of the Department of Medical Physics, MSKCC, and professor of radiology (Physics) Weill Medical College of Cornell University. Between then and Ling's arrival in June, Fuks was acting chair, with monthly visits by Ling for consultation. During those intervening months several important events occurred.

Scanditronix MM50

Prior to Ling's visit to MSKCC in February 1989, Fuks called and said that Ling was to bring a passport for a visit to Sweden. The task was to assess the status of the MM50 (Microtron with a 25 MV scanning photon beam and a 50 MV scanning electron beam) as Memorial was deciding on its purchase.

While in New York before Ling's Sweden trip, Marks invited Ling to play tennis. During the car ride to the tennis complex in Queens, Marks asked whether the MM50 was going to work, and Ling replied that if the MM50 came to New York, it would work.

Actually, Ling had a strategy in mind when he responded to Marks' question. During the visit to Uppsala, Gerald Kutcher, Ph.D., Radhe Mohan, Ph.D., and Ling toured the Scanditronix facilities and viewed the "MSK MM50" being assembled. At the end of the day, the Scanditronix hosts brought out champagne as if to seal the deal. The champagne was accepted graciously, but no commitment was made as to the purchase of the machine.

After their returned to New York, the Memorial team and their radiation oncology colleagues conferred extensively. On the basis of Ling's previous experience with the Varian Clinac 35 at MGH, a strategy was developed involving well-defined payment terms based on the machine's performance criteria. Of utmost importance was an extremely stringent acceptance testing protocol to be performed in Sweden prior to the authorization of shipment of the MM50 to New York. Performing acceptance testing in Sweden was crucial, as it stood to reason that if the machine worked well in Uppsala it should work in New York. In fact, the machine failed the test criteria during the first trial in Sweden, and the Memorial test team had to wait another six months before a second attempt was met with success, and approval granted for the shipment of the MM50 to New York. Chen Chui, Gig Mageras, and Mary-Ellen Masterson jointly worked on the MM50 to implement computer-control radiotherapy.

The MM50 was the most advanced radiation delivery system at the time, and its acquisition was very important for MSKCC to become a leader in radiation oncology and radiotherapy physics. Its important features included the first multileaf collimators, plus scanning electron and photon beams, all under computer control. In addition, the central computer system also controlled gantry and couch motion, thus making it possible to deliver computer-controlled 3D conformal radiotherapy. The computer-control aspect of the MM50 was demonstrated successfully during an NCI site visit and was instrumental in the funding of a large Program Project grant on 3D-CRT from the National Cancer Institute (NCI). The importance of the Program Project grant on 3D-CRT from NCI cannot be overstated as it propelled our research program with financial resources beyond that from the institution. In addition, the MM50 most likely motivated other radiotherapy equipment manufacturers to develop similar advanced features.

Medical Physics Reorganized and Streamlined: New Services

In 1988, the medical physics department consisted of 12 sections, with each section head reporting directly to the Acting Chairman. To streamline the department's governance, Ling recommended the creation of three new services: Clinical Physics, Imaging and Spectroscopic Physics, and Medical Physics Computing[*].

The recommendations were forwarded by Physician-in-Chief Vince DeVita and Marks for approval by the MSK Board. This led to the appointment of Kutcher, Jason Koutcher, M.D., Ph.D., and Mohan, as chiefs of the three new services. Later on, Ron Finn, Ph.D., replaced Larson as chief of the Radiochemistry Service.

Within the Clinical Physics service were the sections of Brachytherapy headed by Lowell Anderson, External Beam Treatment Planning headed by Kutcher, and Calibration/Dosimetry headed by Chen Chui, Ph.D.

The Imaging and Spectroscopic Physics Service comprised Nuclear Imaging and Spectroscopy Physics (Martin Graham, Ph.D., as chief), Magnetic Resonance Imaging and Spectroscopy Physics (Jason Koutcher, M.D., Ph.D., as chief), and Diagnostic Radiological Physics (Lawrence Rothenberg, Ph.D., as chief).

Aside from these services, there were three other sections within the department with center-wide responsibilities, providing essential services for the entire institution. These were Radiation Safety headed by Jean St. Germain, Biomedical and Radiological Electronics with Otis Carpenter as head (assisted by Saul Miodownic in Clinical Engineering and Chester Mah in Radiological Engineering), and Mechanical Instrumentation headed by Joseph Thomas (assisted by Karl Pfaff). These sections offered talent and expertise essential for many new projects.

The purpose of the above, particularly the creating of new services, was to provide a framework for better coordination of activities within each service, and a reporting structure that was more efficient for the operation of the department. Weekly staff meetings were held that included all the service chiefs and section heads. In addition, there were periodical meetings with the Chair and the service chiefs so as to coordinate inter-service collaborations.

Strategic Planning (1990)

In response to the institution's Strategic Planning Initiatives, Ling submitted a report "Evaluation and Strategic Planning" to Vincent DeVita,

---

[*] The Radiochemistry Service was created previously, with Steve Larson, M.D., appointed as Service Chief within the Department of Medical Physics.

M.D., the Physician-in-Chief. The report described both short-term and long-term objectives, with the following major goals:

1. Enhance scholarly and academic excellence.

2. Increase extramural research funding, including

    a. program project grant,
    b. investigator-initiated R01 grants, and
    c. NIH training grant.

3. Collaboratively research with Radiation Oncology on 3 Dimension Conformal Radiotherapy and computer-controlled treatment delivery.

4. Expand research using nuclear magnetic resonance and nuclear imaging.

5. Collaborate with the Medical Imaging Department on development of PACS (picture archiving computer system) and in Positron Emission Tomography.

6. Expand the medical physics post-doctoral training program, and extending the total training period from two to three years.

1995 Leadership Evaluation

Following requests from the Board of Managers for evaluations of leadership performance, each department chair was asked to submit a self-evaluation in areas of leadership, patient care, research, education, program development, and administration. In his submission to Joseph V. Simone, M.D., Physician-in-Chief, Ling addressed his evaluation relative to the goals set forth in the 1990 Strategic Planning document.

Among the highlights, Ling noted that turnover in faculty resulted in the addition of eight new members: Douglas Ballon, Ph.D., Farhad Daghighian, Ph.D., Michael Lovelock, Ph.D., Michael Jackson, Ph.D., Yeh-chi Lo, Ph.D., George Sgouros, Ph.D., John Humm, Ph.D., and Peter Kijewski, Ph.D., significantly enhancing the academic excellence of the department. Ballon, Lo, and Sgouros subsequently became leaders in other programs, while Lovelock, Jackson, Humm, and Kijewski were still at MSKCC in 2010. In terms of patient care, the department functioned at a high level.

The years from 1990 to 1995 saw significant enhancement in research funding and productivity. Extramural research support increased five-fold with ~$1.5 million in direct cost[*]. The research portfolio included an NCI

---

[*]Direct cost is the sum of money that is directly available to the researchers of the research grant. Another sum of money, the so-called indirect cost, is awarded to the institution for functions that support the research.

project grant for three-dimensional conformal radiotherapy (in collaboration with radiation oncology) and individual grants in imaging. In addition, an NIH training grant entitled Training in Radiation Oncology Sciences was obtained to bolster the post-doctoral training program.

The NCI Program Project grant consisted of three research projects and three supportive cores. The first was a clinical project led by Fuks and Leibel on dose escalation for the treatment of cancers of the prostate and lung. The second aimed to investigate new approaches to computer-aided optimization of 3D treatment planning to develop biophysical models of tumor control probabilities (TCP) and normal tissue complication probabilities (NTCP), and to incorporate these biological indices in plan evaluation. Mohan and Kutcher jointly led project #2. The third project, involving Kutcher and Chui, studied the dosimetry of multileaf collimators and incorporated treatment uncertainties into plan evaluation. The three cores were Administration (Ling), Clinical Physics (Kutcher), and Computing/Dosimetry (Mohan/Chui). This program project grant was instrumental to the development of IMRT, as will be described. NCI support for this project continued for three five-year funding cycles, until about 2010.

Other highlights included the installations of CT simulators, the initial development of PACS, and institutional involvement in the Memorial Hospital Credential Committee, of which Ling subsequently became Chair, perhaps the first time a non-physician served in that capacity.

Biophysical Modeling

In the 1980s, the availability of computerized tomography (CT) imaging and increased computer power made it possible to generate radiation therapy treatment plans that displayed the dose distribution overlaid on patient-specific anatomy of both tumors and normal structures. It was then possible to derive "dose-volume histograms" (DVH) of tumors and normal organs that might be related to treatment outcomes, both in terms of tumor control and normal tissue complications. The MSKCC Medical Physics Computer Service, led by Radhe Mohan, Ph.D., with substantial encouragement from the Chair of Radiation Oncology, Zvi Fuks, M.D., developed an in-house treatment planning system with such "three-dimensional computerized radiation therapy" (3DCRT) capabilities. It was put it into clinical use in the mid 1980s.

However, predicting treatment outcomes from individualized dose distributions requires a quantitative understanding of the radiobiological responses of normal tissues and tumors. Simultaneous with the hardware/software developments that made 3DCRT possible, the National Cancer Institute (NCI) funded a Collaborative Working Group (CWG) for an

"Evaluation of high energy photon external beam treatment planning." Four institutions were involved in this project: MSKCC, Massachusetts General Hospital, the University of Pennsylvania, and the Mallinckrodt Institute of Radiation. The CWG results were published in a special issue of the *International Journal of Radiation Oncology, Biology and Physics*—the "Red Journal"—Vol. 21, 1991. Data on normal tissue tolerance, gleaned from the literature and the clinical experience of the senior CWG physicians and physicists, was presented in what is often called "The Emami Paper." There was evidence from clinical outcomes that a 'volume effect' is associated with normal tissue complications, i.e., if a small portion of an organ is irradiated, a higher dose is required to produce the same complication level than if a large portion is irradiated. The Emami paper presented estimates of doses to generate 5% and 50% complication rates for uniform irradiation of 1/3, 2/3, and the whole organ. Also in the CWG volume, MSK's Chandra Burman, Ph.D., and Gerald Kutcher, Ph.D., (together with Michael Goitein, Ph.D., and B. Emami, M.D.) fitted the data of the Emami paper to the "Lyman Model," a three-parameter function that generated complication-specific sigmoidal curves of normal tissue complications probability (NTCP) for specific partial organ irradiation.

The complex dose distributions produced by 3DCRT may differ considerably from uniform partial organ irradiation. In 1989, Kutcher and Burman had developed a method to derive an 'effective volume' for a non-uniformly irradiated organ, which could be used in the Lyman model to calculate NTCP. Their method was elaborated in the CWG issue by four MSKCC physicists (Kutcher, Burman, L. Brewster, and R. Mohan) together with Goitein, and extended to compare the predicted NTCP of several example treatment plans for organs with weak, medium, and strong volume dependences. In the same issue, Burman, Kutcher, Emami, and Goitein provided estimated parameters for the Lyman model for 29 clinical complication endpoints. Some of these estimates are still in use today. In 1992, the MSKCC group (including Mohan, Kutcher, Burman, Gikas Mageras, Ph.D., C. C. Ling, Ph.D., Zvi Fuks, M.D., and Steven Leibel, M.D.) published an alternative histogram reduction technique, the 'effective dose' method, which calculates a uniform dose to the entire organ that causes the same complication rate as the nonuniform dose. Both histogram reduction methods have been applied clinically at MSKCC and elsewhere. In particular, an MSKCC Phase I dose escalation protocol headed by Kenneth Rosenzweig, M.D., with physics support by Ellen Yorke, Ph.D., Andrew Jackson, Ph.D., and Ling, treated inoperable non-small-cell lung cancer with 3DCRT using a limit on predicted lung toxicity, as calculated in the MSK planning system using the effective dose method.

The Lyman model, with its associated histogram reduction methods, is a useful phenomenological expression for comparing rival plans in terms of their potential for normal tissue damage. But a mechanistic model would be more intellectually satisfying and perhaps more clinically useful, allowing extrapolation beyond current clinical scenarios. In 1988, Withers et al. proposed that tissues were made up of single or multicellular functional subunits (FSUs) which were organized in different ways to carry out organ function. Specifically, the subunits of some organs are organized in a parallel architecture (e.g., liver and lung) while those of others (e.g., spinal cord) in a serial architecture. While the radiation response of an individual FS is determined by its basic radiobiological parameters, the organization/architecture of the multiple FSUs that comprise the organ significantly influences the volume dependence of organ complication. Kutcher and Ling encouraged Yorke and Jackson (at that time, respectively, a visitor on sabbatical and a post-doctoral fellow) to study the 'parallel architecture' model in relation to complications with a strong volume effect. In this model, no complication is seen if more than a critical number (a 'functional reserve") of FSUs survives radiation. The MSK group recognized that a population distribution of functional reserves was needed to prevent an overly sharp dependence of complication on dose for organs like lung, liver, and kidney, where there are many thousands of structures that are likely to be FSUs.

From 1990–1992, Jackson created a comprehensive mathematical framework for the parallel model that could be applied to a general dose distribution. As limiting cases, this model also described the dose responses of complications in serial tissues and of the control probability in tumors. About that time, Yorke compared a population averaged parallel model to the Emami tolerance estimates for parallel organs (e.g., lungs, liver, and kidneys) subjected to partial volume irradiation. The comprehensive model was implemented in the MSK planning system and, with parameters suitable for radiation pneumonitis, was used as a clinical constraint in the lung dose escalation protocol, where it had predictive significance superior to the Lyman model.

In 1993, in collaboration with the University of Michigan, Jackson analyzed the incidence of radiation hepatitis in patients treated with 3DCRT for tumors in the liver. Jackson's study was pioneering in several respects:

1. It was the first use of the parallel model to describe complications arising from inhomogeneous irradiation.

2. It was the first analysis based on patient-specific outcome and DVH data.

3. To achieve this, it developed the statistical formalism for applying the maximum likelihood method to analyze DVH and outcome data.

The end results of this analysis were descriptions of the liver's functional reserve and its variation in the patient population, and the dose response of the liver's functional subunits, which could be used to predict the probability of radiation-induced-liver disease (RILD) of future patients subject to inhomogeneous irradiation of the liver. The methods developed in this paper have formed the basis of all subsequent modeling of patient-specific outcomes using DVHs.

Using the increasingly available and powerful imaging and computer tools, MSKCC's radiation oncologists were able to safely increase tumor doses and thus improve the Tumor Control Probability (TCP) factor in the 'therapeutic ratio.' Prostate cancer, a common cancer for which both external beam radiation therapy and brachytherapy are frequently used curative treatments, was a particularly favorable disease for dose escalation. In 1991, an analysis of long-term tumor control outcomes in 679 patients treated with brachytherapy by Fuks, Leibel, and MSKCC surgeons found that patients whose disease in the prostate was eradicated experienced a significantly lower incidence of distant metastases. Significant reduction of distant metastases in biopsy-proven locally controlled prostate cancer patients treated with external beam was later reported by MSK radiation oncologist Michael Zelefsky, M.D., and others. For prostate cancer patients, local relapse-free survival (LRFS) is a significant predictor of long-term distant metastases-free survival (DMFS), a point of particular impact because localized disease is seldom a cause of mortality. Similar correlations have been reported for other types of cancer. In 1993, Yorke, Ling, Fuks, Larry Norton, M.D., and Willet Whitmore, M.D., developed a mechanistic mathematical model of the development of distant metastases from primary and recurrent tumors and fit its parameters to the brachytherapy data.

Further pioneering studies followed, developing techniques to analyze the dependence of radiotherapy outcome on the patient-specific dose distributions, generated by 3DCRT and IMRT. The tolerance doses generated by these studies were incorporated into treatment planning procedures to ensure the safety of patients treated. In 1999, Jackson and Mark Skwarchuck analyzed the dependence of rectal bleeding on dose volume histogram variables in prostate cancer patients treated with 3DCRT. They were the first to use permutation tests for calculation of p-values; area under the curve of the receiver operator characteristic for assessing predictive power of models; multivariate modeling, including both DVH, dose distribution geometry, and clinical variables; and comprehensive testing of the full range of DVH variables. The multivariate model showed that the incidence of rectal bleeding was correlated with the following factors: the volume of rectum receiving doses >70 Gy, the surrounding of the rectal wall by the 50 Gy isodose line, the absolute volume of the anatomic rectal wall; and the patient's age and diabetic status. Dose volume constraints from this study were implemented in treatment planning of patients treated for prostate cancer and are still used today.

In 2000, Sabine Levegrun, Ph.D., and Jackson analyzed biopsy outcome in patients treated with 3DCRT and IMRT for prostate cancer. An initial paper used DVH variables and found that mean dose to the PTV, but not minimum dose, correlated with biopsy outcome. A second paper used TCP models, the first to use patient-specific outcome and dosimetric variables, and found that only particular combinations of the model parameters could be determined from the clinical data. A third paper used multivariate logistic models based on patient risk categories and tumor mean dose, showing that dose responses of the different risk categories were significantly different. This was the first study to fit a multivariate model of tumor dose response to patient specific clinical data and the first to use bootstrap Monte-Carlo techniques to assess uncertainties in the modeling process.

The Phase I dose escalation trial for patients treated with radiotherapy for NSCLC, conducted during the 1990s and into the 2000s at MSKCC, made use of the models of radiation pneumonitis to determine patient-specific prescriptions. An initial LKB NTCP limit of 20% was later increased to 25% based on clinical data. In its latter stages, escalation was conducted independently in cohorts stratified by a fractional damage based on a parallel model. Under the guidance of Kenneth Rosenzweig, M.D., Yorke, and Jackson, this trial successfully treated patients to 80 to 90 Gy. Dose volume factors correlating with the incidence of radiation pneumonitis in this patient cohort were analyzed by Jackson and Yorke.

From the mid-1990s onwards, after implementation of IMRT and dose volume constraints derived from our analysis of toxicity, the incidence of rectal toxicity in patients treated in our Phase I dose escalation trial for prostate cancer decreased dramatically. Analysis of patients treated during the latter stages of this trial by Joseph Bauer, Ph.D., and Jackson in 2006 was unable to find convincing dose volume factors affecting rectal bleeding, probably because of the low rate of bleeding (~2 to 4%), though a later study of dose distribution geometry by Reshma Munbodh, Ph.D., and Jackson in 2008 showed some indication that irradiation of the upper part of the rectum was more toxic.

In collaboration with Suzanne Wolden, M.D., in 2005, Jackson studied the incidence of hearing loss resulting from MSKCC's 3DCRT and IMRT of head and neck cancer showing that the incidence of clinically significant hearing loss increased with mean cochlea doses >45 Gy, and with the number of cycles of cisplatin chemotherapy.

With support from both ASTRO and AAPM in 2007, the first major review and synthesis of published dose volume dependencies of tolerance doses since the Emami paper was initiated, called QUANTEC (Quantitative Analysis of Normal Tissue Effects in the Clinic). Both Jackson and Yorke participated as authors and members of the steering committee. In 2010, the project resulted in a special issue of the Red Journal, including 16 papers on toxicity in specific organs, together with a summary table and reporting recommendations. The QUANTEC issue synthesized the most important developments in the field of toxicity modeling over two decades.

Shortly after the completion of the QUANTEC project in 2010, Jackson, as head of AAPM's Biological Effect sub-committee, founded two projects aimed at covering areas not dealt with by QUANTEC. Through an AAPM working group, in collaboration with Jimm Grimm, Ph.D., and Larry Marks, M.D., Jackson co-founded the HyTEC project (which aims to synthesize tolerances and tumor control for hypofractionated SBRT treatments), and appointed Grimm and Yorke as co-chairs of the group. Jackson and Yorke also serve as members of the HyTEC steering committee. In collaboration with Louis S. Constine, M.D., he co-founded the PENTEC project (which aims to synthesize tolerances for pediatric treatments), led the outcome modeling group and served with Yorke as member of the steering committee.

The major limiting factor for QUANTEC was the poor standard of reporting in the literature. To improve upon this, Jackson, Yorke, and Rosenzweig proposed a comprehensive reporting method called the dose-volume atlas of complication incidence, which has proved useful in the recent efforts undertaken by the HyTEC group. With the advent of electronic supplements, other options—such as reporting individual DVHs and associated outcomes and clinical variables—have become viable.

Through the 2000s and to the present, SBRT treatments of small tumors have grown in prevalence, and studies of the outcomes of these treatments at MSKCC overseen by Jackson and others have provided tolerances for chest-wall pain after treatment for early-stage NSCLC (with Fan Liu, Ph.D., Eric Williams, Ph.D., Yorke, and Andreas Rimner, M.D.), severe esophagitis after treatment of paraspinal metastases (with Brett Cox, M.D.), and brachial plexopathy after treatment for either early-stage NSCLC or paraspinal metastases (with Williams, Yorke, and Rimner).

In collaboration with Nancy Lee, M.D., Jackson studied the incidence of dysphagia in patients treated with IMRT and chemotherapy for head and neck cancers. They found that the incidence of both grade 2 and grade 3 dysphagia was dependent on the mean radiation dose to the total constrictor muscle volume, the disease site, and patient age. Limiting the total volume of constrictor muscle to <58 Gy was shown to keep the predicted rate of G2 dysphagia to <5%.

Development of IMRT at Memorial Hospital

In the mid-1980s, an important paper by Anders Brahme, Ph.D., at the Karolinska, described the basic idea underlying what is now called intensity-modulated radiation therapy (IMRT) that yields a dose distribution that conforms to the target volume, including concavities with adjacent normal structures. This paper stimulated many groups to study various aspects of this method, namely treatment planning using the inverse method and delivery of intensity-modulated beams.

In the spring of 1992, Jerry Kutcher returned from a lecture and visit to MD Anderson. He reported to Ling that he met Thomas Bortfeld, a post-doc of Arthur Boyer from Heidelberg, who had developed an inverse planning algorithm for his Ph.D. thesis. At that time in 1992, Bortfeld and colleagues had just demonstrated a successful delivery of inversely planned IMRT to a phantom using multileaf collimators at MD Anderson. Kutcher suggested that Bortfeld be invited to New York en route to his return to Germany that summer. Ling and Mohan subsequently invited Bortfeld to spend a week or two in New York to see whether his algorithm could be linked to the MSKCC treatment planning system.

During Bortfeld's visit, Radhe Mohan and his post-doctoral fellow, Xiao-Hong Wang, worked with Bortfeld to adapt his inverse planning algorithm to the Memorial 3D-CRT planning system. Subsequently, Wang and Mohan modified and improved the algorithm. Their joint effort created a viable inverse planning module that was integrated into the treatment planning platform in clinical use at Memorial.

Concurrently in the early 1990s, a graduate student at Columbia University, Spiridon Spirou, applied to Chen Chui, who became Spirou's mentor for his Ph.D. thesis project. Their collaborative research resulted in the so-called sliding window technique for delivering an intensity-modulated radiation beam using multileaf collimators in the dynamic mode and also an inverse planning algorithm. The development of the various components for IMRT is discussed in more detail in chapter 8 on Computer-Assisted Treatment.

The dual abilities to perform inverse planning and to deliver intensity-modulated beams set the stage for implementing IMRT at Memorial. Unfortunately, the existing linacs were unable to deliver radiation while the multileaf collimator (MLC) leaves were moving. A serendipitous rainstorm saved the day.

Ling was on his way to attend the June 1993 International Conference on Radiation Oncology in Kyoto. It was raining heavily as he looked for a cab, but Tim Guertin and Marty Kandes of Varian came to his rescue with a car. During the trip to their respective hotels, they discussed how DMLC and IMRT could work. Ling was aware of Varian's so-called Kyoto software for delivering conformal arc therapy in which multileaf collimator leaf positions were controlled by the gantry angle. He suggested that the MLC leaf positions could be controlled by the monitor units (MU) instead of the gantry angle, thereby performing DMLC for IMRT.

In October 1995, more than two years after the Kyoto meeting, the first IMRT patient was treated at Memorial. During those two years, important advances included: the work of Radhe Mohan's group in the inverse planning algorithm, Chen Chui and Spirou's work in DMLC leaf sequencing and dose calculation, LoSasso's contribution to the physical/dosimetric aspects of DMLC in IMRT, Wang's work in IMRT dosimetric verification, Burman's QA methodology, and Kutcher's overall guidance in the clinical implementation of IMRT.

At the same time, Fuks and Leibel, heavily involved as radiation oncologists, spearheaded the clinical aspects of the IMRT program. Finally, and importantly, Varian provided the control systems for using DMLC for IMRT.

Fuks recollected that because there was no method of visualizing the treatment fields delivered by DMLC, physicians and physicists found it difficult to check and sign off each treatment field prior to delivery. He indicated that "the transformation into trusting the computer record and verifying system for DMLC function constituted a psychological barrier that had to be overcome." Furthermore, in his words, the "first delivery of DMLC-medicated IMRT constituted an act of crossing the Rubicon that psychologically initiated a process that facilitated image-guided radiation therapy (IGRT)." He added, "Since then, we have adopted the routine of on-line dynamic treatment procedures totally controlled by computers."

The method for treating the first IMRT patient at Memorial used the Bortfeld/Mohan/Wang inverse planning algorithm to calculate the IM beam profiles, and the Spirou/Chui DMLC algorithm to deliver them. Subsequently, after much validation/verification, the Chui/Spirou inverse planning algorithm was integrated into the Memorial planning system for routine and large-scale clinical implementation.

IMRT ushered in a new era in radiotherapy. Because of its ability to "dose-paint" or "dose-sculpt," IMRT can minimize dose deposition to organs-at-risk abutting the tumor. This ability permitted dose-escalation to the tumor, which was hypothesized to increase local tumor control. This hypothesis was tested in clinical trials conducted at MSKCC and elsewhere, initially in prostate cancer and then extended to other disease sites. At MSKCC, the dose delivered to the prostate cancer was escalated to 76, 81, and eventually to 86 Gy using IMRT. Fuks, Leibel, and Zelefsky subsequently reported improved outcome when higher doses were employed.

The IMRT technique using inverse planning and dynamic MLC, pioneered at MSKCC, was widely adopted by other centers within a few years as commercial planning software became available. Subsequently the technique evolved into VMAT (volumetric modulated arc therapy), which combines gantry rotation and dynamic MLC to increase the speed of IMRT delivery. IMRT and VMAT are the standards of radiotherapy today.

IMRT QA Procedures

Multileaf collimators were introduced at MSK in the early 1990s as a replacement for cerrobend blocks to conform radiation to tumors while shielding normal tissues. This principle was demonstrated at MSK with the Scanditronix MM50 racetrack microtron with a multileaf collimator acting as the upper collimation jaw. Then Varian made available a tertiary multileaf collimator mounted below the lower collimator jaw. Soon thereafter, decades of striving for homogeneous dose distributions using uniformly flat beams was upended by the seemingly preposterous idea to intentionally deliver intensity-modulated (non-uniform) beams to tumors with the MLC system. To implement this new technology, referred to as intensity-modulated radiation therapy (IMRT), using the sliding window or dynamic multileaf collimation (DMLC) approach, treatment planning software development, dosimetric testing, and quality assurance procedures were carried out by MSK medical physicists over a three-year period before the first IMRT treatment was used for a patient with prostate cancer in October, 1995. This new treatment method presented dose planning, delivery, and verification hurdles inherent in the use of numerous small radiation subfields optimized with inverse planning techniques developed at MSK.

Calculation algorithms at MSK in the early 1990s used a pencil beam convolution, whereby open-field-tabulated data was modified by a ratio of convolutions of MLC-shaped fields and open fields, coupled with a multicomponent source function. Leaf sequencing algorithms, constrained by both maximum leaf speed and field size, emphasized the 3D impact of the MLC configuration. Three physical MLC parameters, significant to IMRT-specific dose delivery, were identified and modeled for each photon energy: average of mid-leaf and interleaf transmissions, head scatter, and leaf end transmission. Incremental improvements in dose calculation algorithms took into account all these factors.

The consequences of leaf misalignment due to calibration error or mechanical drift are clearly much more important for IMRT, because leaf position errors affect the dose throughout the target volume; whereas for 3D CRT with static MLC the leaves only shape the field edges. Responsibility for accurate dose delivery has always been shared by the manufacturer and MSK medical physicists. Agreements between primary and secondary position readouts and between primary readouts and leaf sequence file data are continuously monitored by Varian software in real time with beam interlock capabilities. Off-line, leaf positions are monitored using the so-called "picket fence" radiographic analysis, and the dose output for reference IMRT fields at different gantry and collimator angles are measured. The results of these QA procedures has been be acquired over the last 20 years, with increment improvements in both hardware and software. As we gained confidence, the original semi-weekly QA testing has been reduced to monthly intervals.

From its inception, IMRT raised the anxiety level of medical physicists because of concerns for potentially serious dosimetric errors, resulting from software or human errors. As compared to treatment with static fields, monitor units for IMRT are not simple to predict, and verification images do not depict the intensity modulation within the treatment fields. These concerns rightfully raised the bar when designing appropriate QA procedures. Therefore, for the first ~400 IMRT patients, all treated for prostate cancer, the dose calculated at the isocenter was verified with point dose measurements in a homogeneous phantom for every IMRT field prior to treatment. The relatively complex modulation encountered with head and neck sparked an upsurge in 2D dosimetric measurements with film, and subsequently with the electronic portal imaging device (EPID). Current approaches to patient-specific treatment QA at MSK rely heavily on more efficient computer verification at multiple stages in the planning and treatment process.

Arrival of PACS (Picture Archive and Communication System)

One of the new faculty positions, negotiated in 1988 during Ling's MSKCC recruitment, was for the development of PACS (Picture Archive and Communication System). Upon Ling's arrival at MSKCC, Robin Watson, M.D., then Chair of Medical Imaging, became ill, and recruitment for the PACS position was postponed. After Ronald Castellino, M.D., became Chair, and the name of his department changed to the Department of Radiology, the PACS faculty position was created and included in the operating budget for 1993. This was accomplished despite the reluctance of the then associate hospital administrator for Medical Physics, but fortunately Dr. Paul Marks remembered his commitment to that faculty position and authorized it to be included in the 1993 operation budget.

The Clinton health-care reform, though stillborn, prompted budget constraint at MSKCC and an across-the-board 10% reduction in the 1994 operating budgets for all departments. As the PACS position was still unfilled at the budget submission date (autumn of 1993), it was an easy decision to eliminate it from the budget, at least temporarily. The position was reinstated subsequently, and Peter Kijewski, Ph.D., was recruited to start in this position in 1994.

Kijewski and Ling had known each other since the mid 1970s when they were at Harvard's Joint Center for Radiation Therapy (JCRT) and MGH, respectively. Kijewski, with Sam Hellman, M.D., and Bengt Bjarngard, Ph.D., was pioneering dynamic radiotherapy in the 1970s at JCRT, and he was well known as an expert in computerized treatment planning. Although Kijewski had no prior experience with PACS per se, Ling believed that given Kijewski's extensive experience with computer systems, particularly in the use of images for treatment planning, he could readily develop an expertise in PACS, and that proved to be true.

After joining MSKCC, Kijewski began exploratory work with radiologists in developing PACS. In 1995, a phase 1 configuration was installed that included image acquisition from three CTs and one MRI scanner, an image storage server, a link to Radiology Information System (RIS), and image distribution to a dozen viewing stations with single- or dual-monitor screens. George Bosl, M.D., Chair of the Department of Medicine, and Jerome Posner, M.D., Chair of the Department of Neurology, agreed to perform an initial evaluation of the potential benefits of PACS for clinical services. Both Bosl and Posner were extremely enthusiastic about the performance of the prototype PACS system and supported its continued development.

Samuel J. Dwyer, III, Ph.D., a PACS expert with extensive practical experience, was invited as the J. Laughlin Visiting Professor May 29–30, 1997. In addition to his presentation of the John S. Laughlin Lecture, Dr. Dwyer led multiple workshops to share his PACS experience with Memorial staff. (The agenda of his visit is on p. 46 in the section describing the John Laughlin Visiting Professors.)

During the planning phase for the new outpatient facility at 53rd Street in 1997–1998, hospital administration requested that the radiology operation at that new site be filmless. The Department of Radiology (Drs. Castellino and Larry Schwartz) and Medical Physics (Kijewski and Ling) conducted an extensive investigation of available commercial systems. The decision process was difficult as a fully developed functional and integrated (both front-end and back-end) system was not immediately available. Eventually, with Herculean collaborative efforts by the GE and Memorial teams working through the weekend prior to going live, PACS became a reality in December 1998.

Before the startup of PACS, prior x-ray images on film were digitized for importing into PACS. CT and MRI exams stored on MOD (multiplex optical data storage) were also imported into PACS. This was done primarily in preparation for the opening of the 53rd St. facility so that previously acquired exams could be reviewed. The Memorial implementation represented the first to use PCs in both the front and back ends and Ethernet to link the main campus and the 53rd St. facility.

Initial use of PACS was by selected Radiology services, with a phased expansion to additional Radiology services. This was followed eventually by the opening of a filmless facility at 53rd St. in June 1999. Clinicians reviewed images on film on Friday prior to transferring to the new facility, and transitioned to 100% review of images on PACS the following Monday. In spite of this abrupt transition, PACS received immediate wide and appreciative acceptance by the clinical staff.

Growing Pains

Starting in the mid-1980s, it became obvious that expansion both within and beyond the limits of the main campus was going to be essential. Many inpatient procedures were now converted to outpatient status, and more outpatient space could not be accommodated on the main-campus block. Thus began the growing pains with the relocation of various programs. The discussion below pertains only to the expansions that involved radiological and radiation oncology physics services.

MSK Expansion Involving Diagnostic Physics

1. *The 64th Street Breast Center.* This center was opened to relieve the need for mammography procedures and chemotherapy for breast cancer patients at the main campus. A reduced presence for mammography procedures remained at the main campus, with the bulk of these procedures going to the 64th Street Center.

2. *The Evelyn Lauder Breast and Imaging Center.* The increasing number of breast cancer patients referred for evaluation and treatment outgrew the 64th St. location and a new building, the Evelyn Lauder Breast and Imaging Center (BAIC), was built on the block between 64th and 65th streets on Second Avenue. The mammogra-

phy programs previously at the 64th St. site and the Guttman Clinic were fused with the BAIC programs and expanded.

3. *MSK 53rd Street Outpatient Facility.* This facility, opened in 1998, represented a significant expansion of the clinical activities for outpatient visits, including radiology examinations with CT, radiography, and mammography. Diagnostic Radiological Physics provided services and, as described previously, PACS was implemented for a filmless environment for the first time.

4. *The Sidney Kimmel Center for Prostate and Urologic Cancers.* This center was opened to provide outpatient services for the departments of surgery and medicine seeing outpatients with prostate and urologic cancers. A dedicated urological radiographic procedure room was included at the center.

Other sites supported by the radiological physics group included MSK Guttman in Greenwich Village, the Beekman-Downtown Hospital (Lower Manhattan), and the Breast Examination Center of Harlem (BECH).

The medical physics support for the above imaging centers were provided by Rothenberg and Rick Fleischman, BS.

MSK Expansion Involving Radiation Oncology Physics

To serve the patients in the metropolitan areas outside of the five boroughs, the Regional Care Network of radiotherapy was created. The network extends state-of-the-art radiotherapy technology and a multimodality approach to those who need cancer care, and it obviates the inconvenient travel to Manhattan during patients' treatment courses. These regional centers within the network bring the same high level of expertise, clinical care, and access to clinical trial protocols that are available at the main campus. At the time of this writing, there are four regional centers that provide radiation oncology services located within 40 miles of the main campus. Two sites are in Long Island, at Commack and Rockville Center; one site is in Westchester County in Sleepy Hollow; and one site is in New Jersey in Basking Ridge.

In addition to the radiation oncology physics efforts at these sites, diagnostic radiological physics support was also provided. With the recommendation from Kutcher, Chandra Burman was appointed to oversee the clinical physics of the regional center and to ensure that the same standards and practices were implemented throughout the network.

1. *MSK at Phelps Memorial Hospital, Sleepy Hollow, 1995.* Memorial Sloan Kettering opened its first regional outpatient treatment facility in October 1995 in Sleepy Hollow to offer chemotherapy. It is

located at the Phelps Memorial Hospital. Radiation oncology services became available there in January 1997. C. Obcemea, Ph.D., was in charge of medical physics at Phelps.

2. *MSK at Saint Clare's Hospital, New Jersey, 1996.* In June 1996, Memorial Sloan Kettering opened an outpatient program at Saint Clare's Hospital in New Jersey, with Joseph Hanley, Ph.D., as chief medical physicist. Ten years later, the program was moved in 2006 to the new ambulatory care center in Basking Ridge, New Jersey. Maria Chan, Ph.D., was the chief physicist at this site.

3. *MSK at Mercy Medical Center, Rockville Center, 1997.* In November 1997, Memorial Sloan Kettering opened the ambulatory care facility at Mercy Medical Center in Rockville Center on Long Island to offer chemotherapy and radiation treatment in Nassau, Western Suffolk, and Queens counties. David Huang, Ph.D., managed the medical physics activities there.

4. *MSK in Commack, New York, 2002.* Memorial Sloan Kettering opened its first freestanding outpatient treatment facility in Commack, New York, in June of 2002 to offer a range of services, including cancer diagnosis, chemotherapy, and radiation therapy. Ruemei M.A., Ph.D. was chief physicist at this site.

1997—Evaluation of the Department and Strategic Planning

In this document, submitted to Physician-in-Chief David Goldie, M.D., Ling provided a detailed discussion of each service/section and each of the faculty members. Interestingly, Ling worked for and interacted with five physicians-in-chief and one acting physician-in-chief during his time at Memorial Hospital. These included Hellman during Ling's recruitment, then Drs. DeVita, Thomas Fahey (acting), Joseph Simone, David Goldie, and Robert Wittes.

Of significance was the pending departure of Jerry Kutcher to Cambridge University to pursue a doctorate in the history of science and Lowell Anderson's retirement. Research, extramural funding, and the education programs were identified as strengths. Needs were also identified in specific areas for development as described below.

In Radiation Oncology physics, our centerpiece research program was the PO I grant, funded for a second cycle of five years. Relative to that, plans were being developed for installing a CT scanner (supported by extramural funds) into the treatment room that houses the Scanditronix MM50. By scanning the patient in the treatment position before or after radiation treatment, we shall better understanding the nature and magnitude of setup error and organ motion and their effect on treatment outcome. This project will have significant impact on radiation treatment methods and results in the future.

To improve our treatment planning capabilities, Medical Physics staff were spearheading the implementation of CT-simulation, installing a second Picker AcQSim CT. Subsequent to that, our 3D planning system will be upgraded to the Windows/NT platform so as to be compatible to the other components of the MSKCC Enterprise, e.g., PACS, disease management system (DMS), and other computer systems.

In the strategic planning area, emphasis was placed on the development and deployment of PACS in a step-wise implementation in main Radiology, the 53rd St. site, MSKCC 64th St., and subsequently the other departments, clinical areas, and surgical suites. This ambitious undertaking required the active and well-coordinated participation by Radiology, Medical Physics, Administration, and Hospital Information Systems.

The strategic planning also included plans to increase research in diagnostic physics, e.g., in image reconstruction algorithms, signal/image processing, observer analysis, etc. Another area relates to the physics and computer support for the whole-body PET scanner. For this the recruitment of a medical physicist with expertise in PET was necessary.

This report also provided the first mention of "biological imaging" and its research and clinical potential. This was propelled by the capabilities of magnetic resonance imaging and spectroscopy (MRI and MRS) and of positron emission tomographic imaging (PET). The potential of combining MRI and MRS (of ^{1}H, ^{12}C, ^{19}F, and ^{31}P) can yield a wealth of information on tumor biology, physiology, metabolism, and response to anti-neoplastic therapy. PET imaging may produce additional and complementary information on the above and on pharmacokinetics. An important challenge was to understand the biology underlying the images and spectra, and to devise a strategy to optimize cancer diagnosis management based on such understandings.

In addition, PET will be applied to better quantify the activity and absorbed dose distribution of radiolabeled monoclonal antibodies (MoAb). Such quantification is the first step in the treatment planning of RIT (analogous to the treatment planning of radiotherapy using external beam or implanted sources). The acquisition of a mini-PET device for animal studies was also planned to facilitate laboratory research in this area, complementary to the GE animal MRI/MRS 4.7 T system to provide correlated images from MR and PET.

Goldie participated in the recruitment of Howard Amols, Ph.D., then chief physicist at Columbia University, as the new service chief of Radiation Oncology Physics. To replace Anderson, Howard recommended Marco

Zaider, also from Columbia. Both Howard and Marco contributed significantly in the clinical and academic aspects of the department during their tenure.

Image Guidance for Radiotherapy

Portal Imaging

Traditionally, image guidance at MSKCC and elsewhere was done using portal imaging. The initial setup of the patient was done, as it still is, by aligning skin marks with the room lasers. Then, for image guidance, the MV treatment beam was used to image the patient with x-ray film placed in a cassette downstream of the patient. Comparison of the MV image with a reference kV image taken at the simulator was performed by a physician or therapist, and correction of the patient's position, if needed, was implemented. A MSKCC innovation was the use of a graticule placed into a slot in the head to the treatment machine. The graticule tray, with a reference grid of points, spaced at 2 cm, made possible the alignment of chosen bony features in the portal image with those in the reference image. The process was time consuming; at best it was done weekly, or one out of every five treatments

Development of Volumetric Image Guidance

Lacking in the traditional approach was the ability to image and, therefore, to accurately position most soft issue targets. By 2000, two development projects were underway to avail 3D volumetric imaging for improved setup accuracy. In the first project, MV projection images were acquired as the gantry rotated around the patient to produce MV cone-beam CT (CBCT). The second project involved the installation of a diagnostic CT scanner in a treatment room.

The MV-CBCT Project

The MV CBCT project started in early 2000, as demand increased for more accurate patient setup for Stereotactic Radiosurgery (SRS) and Stereotactic Body Radiotherapy (SBRT). Initially, the group included Drs. Jeng-hwa Chang, Gig Mageras, Howard Amols, and Clifton Ling. Most of them were also involved in the EPID (electronic portal imaging device) dosimetry project in late 1990s, and they had a good understanding of the EPID system. A research fellow, Dr. Eric Ford, was engaged for the project. The aim was to prove the principle and identify potential problems for the clinical use of MV CBCT. By mid-2001, it was evident that MV CBCT could be implemented for 3D imaging of lung tumors. However, the imaging and analysis time was long (~one-half day) and the dose was too high (~200 cGy per scan) for clinical use.

In 2003, another post-doctoral trainee, Jussi Sillanpaa, Ph.D., continued the effort. To reduce the time, the image acquisition and reconstruction processes were automated. From the EPID dosimetry project, we knew that the EPID images were without a record of gantry angle in the image header. Thus Jussi developed a method for determining the gantry angle from MLC leaf positions and record them in the image header. To achieve low-dose MV CBCT, we collaborated with Varian between 2002 and 2006 and carried out the project in three phases. In Phase 1, a specially designed interface box was employed for synchronizing the linac and image acquisition electronics, which led to the reduction of image dose to 0.2 MU/projection.

In Phase 2, we successfully developed two gated CBCT techniques and adopted weighted-expectation maximization reconstruction algorithms to mitigate the image artifacts, which were due to the limited number of projections that were unevenly sampled over the view angle.

During Phase 3, the MV CBCT system was further improved by adopting a high-efficiency image receptor composed of CsI crystals. In addition, a new synchronization circuit was developed, allowing us to limit the exposure to one beam pulse, i.e., 0.028 monitor units (MU) per projection image. With this MV CBCT system, a 2% electron density difference was discernible with a maximum absorbed dose of about 12 cGy per scan. This low-dose MV CBCT was used to measure the inter-fractional variation in lung tumor position for eight patients, with results published in the Red Journal (2007). This was the first patient series using onboard CBCT at MSKCC and one of the first few reported IGRT studies in the world. As a result, this paper was selected by ASTRO in 2007 for on-line CME credit.

Image-Guided Treatment Facility with an In-room CT

During 1999, a group of physicists led by Wendell Lutz, Ph.D., implemented in-room CT-guided treatment. This was done by (1) installing a CT scanner, a Picker Premier Xtra, in the corner of a treatment room; (2) developing a special-purpose stereotactic body frame (SBF) with vertical-to-horizontal transport; and (3) constructing a rail system by which the patient in the stereotactic body frame was transferred from a vertical setup position in the SBF to a horizontal position on the CT couch and, finally, to the linac couch for treatment. The CT scanner made possible 3D localization and, hence, the targeting of the tumor. The rail system was designed such that, with the linac and CT couches rotated to the appropriate angles, the SBF in which the patient was immobilized could be smoothly moved from the CT couch to the linac couch. This facility was designed to study whether or not targets in the thorax and abdomen could be treated with extreme accuracy, that is, positioned to within 1.5 mm for multiple fractions.

The Stereotactic Body Frame

The SBF was developed by Kamil Yenice and Wendell Lutz for use in the image-guided facility. Studies were undertaken to address three questions:

1. Could we rigorously immobilize a patient during the lengthy interval from CT imaging localization to linac treatment delivery?

2. Could we develop a 3D stereotactic coordinate system that would permit very accurate target localization with the CT, which could be translated to very accurate setup positioning for linac treatment?

3. Could the SBF be designed so that patients could be reproducibly positioned accurately enough so that daily CT localization was unnecessary?

The CT scanner was used to localize the patient's anatomic structures in the stereotactic coordinate system of the SBF. The SBF has (i) a visual component: longitudinal and height scales, a movable bridge with laser pointer that spanned the SBF left to right; and (ii) an internal component that could be visualized in the CT. Using a system of wires embedded in the SBF that were arranged in Z patterns, it was possible to localize any object seen in the CT in the stereotactic coordinate system.

In brief, the clinical process was as follows:

1. Patients were set up in the SBF in a vertical position to make it easier for them and hopefully to facilitate more accurate repositioning. A motorized support stand was used to tilt the SBF from a vertical to horizontal orientation. Once horizontal, patients were immobilized by a set of seven pressure plates: three to immobilize the rib cage and four to immobilize the pelvis. The head and body were also cradled in customized expanded foam. Each of the seven pressure plates was indexed in 1 mm steps to enable repositioning on a daily basis.

2. The pretreatment CT scan was acquired. Visualized in the scan was the patient's anatomy and the system of fiducial wires that constituted the stereotactic coordinate system.

3. The stereotactic coordinates of specific anatomic features were recorded.

4. In preparation for the image-guided treatment, the patient was again immobilized in the SBF and scanned using the in-room CT.

5. The stereotactic coordinates of the same features were determined. Any change in the stereotactic coordinates of the features or the setup error could, therefore, be calculated.

6. After transfer to the linac couch using the rail system, the SBF was set up using the external scales and room lasers. Any setup error of the patient in the SBF determined from the prior in-room CT scan could be implemented as an adjustment of the SBF setup coordinates.

7. A portal image check was made.

8. Treatment delivery.

Patients with paraspinal disease were treated in this facility in the years 2000 to 2003. The SBF and the image guidance facility provided us with invaluable clinical experience. Studies (1) and (2) were successfully completed, but daily CT localizations established that the lofty goal (3) could not be achieved with our SBF. Nonetheless, much was learned about patient comfort and intra-fractional motion, resulting in new and improved workflows for image-guided treatments.

Image Guidance and The Development of Spine Stereotactic Body Radiotherapy (SBRT) at MSKCC

Image guidance played a key role in the beginning of SBRT at MSKCC. SBRT, characterized by very high doses delivered in typically one to five fractions, necessitates that the dose be delivered with very high geometric accuracy. At MSKCC, the first disease sites, outside of the brain, using SBRT were metastatic spinal and para-spinal tumors.

One of the lessons learned from the image-guided treatment facility was that even for patients immobilized in the SBF, they may still move after the initial setup. Thus, it is desirable to image the patient immediately prior to treatment and, in case of intra-treatment movement, to be able to quickly re-image the patient after treatment interruption.

The Move to Strict Image Guidance and the MSKCC Paraspinal Immobilization Cradle

At the time the image-guided facility was being developed, an alternate approach was also being explored specifically for the treatment of spinal and para-spinal disease using SBRT. This was possible because many of the patients being treated had had spinal augmentation hardware implanted during surgery, which was clearly visible in MV images and served as excellent fiducial markers. This involved the development of an immobilization cradle in which patients could be set up quickly. The stereotactic approach was abandoned, and the cradle has no coordinate system other than external scales for the initial setup of the patient in the cradle. Certain features of the SBF were retained: four large lateral paddles were used to position hips and the rib cage. Spine patients were always treated "arms down" for comfort.

Immobilization was monitored using an infrared tracking system that was developed in-house. The control software, developed by Ping Wang from the Computer Services Section, was designed with therapist input for an efficient clinical workflow. Two or three small infrared reflective markers were taped to the patient's skin on relatively fixed points, such as the sternum or hips. Stereoscopic tracking cameras mounted in the treatment rooms allowed the therapists to track the motion of the markers throughout the treatment session. The therapists could stop treatment and re-image the patient to check and, if necessary, correct the target position. Five immobilization cradles were made in the instrument shop under the leadership of Karl Pfaff. They were used in the years 2001 to 2013 to deliver over 1000 treatments to para-spinal patients.

MV Image Guidance for Spinal Treatments

In the years 2000 to 2004, prior to the installation of the first gantry-mounted kV CBCT imaging system, spinal patients with spinal hardware were generally positioned using orthogonal 2D-2D MV imaging. At this time, Varian had developed a tool in their 'Portal Vision' software that permitted setup accuracies of ≤ 1 mm. Contours could be drawn on digitally reconstructed radiographs (DRRs) which served as reference images. During the imaging session, just prior to treatment, the DRRs were overlaid directly onto the MV localization image. Using this registration process, the therapists could rapidly and accurately correct the target position by altering the couch position from the console.

Another key development was improvement in the quality of metal visualization in the reference images. Picker CT scanners came with an AcQSim localization workstation. In addition to DRR, a digitally composited radiograph (DCR) could also be produced. With a DCR, the user can select ranges of Hounsfield units (HU) and to each, assign arbitrary opacities. Using a DCR it was possible to create reference images in which spinal hardware was clearly delineated. Furthermore, it became possible to clearly see small implanted gold markers in the reference images. All patients treated using MV image guidance during these years who did not have spinal hardware had six gold markers, two per vertebra, implanted into the posterior elements of their spine, making for robust and accurate target positioning.

The Varian On-Board Imaging System

In July, 2004, the long-awaited kV imaging system was retrofitted to the Varian 21EX linac in the image-guided treatment facility. It featured a powerful kV fluoro unit and a large amorphous silicon detector mounted on the gantry at right angles to the MV beam. In addition to radiographic imaging and fluoroscopy, it had several modes of cone-beam acquisition. Initially, the CBCT modes were unavailable. Peter Munro, the Varian physicist who did the installation and testing, did, however, provide us with a seven-page document that allowed us to acquire a CBCT scan with an 85-step process. Nonetheless, it was eventually used for the first patient CBCT scans. After the CBCT obtained FDA approval, a clinical trial established that the CBCT image guidance of spinal targets without implanted fiducials was at least as accurate as orthogonal MV imaging of implanted markers in spinal targets. CBCT-based setup imaging has also resulted in the widespread adoption of automatic 3D registration, where the therapists select a volume of interest and then use software to perform the match.

This new imaging system simplified the positioning of the target with respect to the radiation beams; it was a strict "image guidance" system; the immobilization systems had no associated stereotactic coordinate frame. Software tools that greatly facilitated the registration of the acquired image with the corresponding reference led to much more accurate determination of the couch corrections required to correctly align the patient. Associated with the introduction of image guidance was the introduction of highly accurate patient support couches, which can be driven directly from the registration results of the imaging procedure. Initial positioning accuracy was typically <1 mm. Subsequent refinement included the ability to pitch and roll the couch, giving all six degrees of freedom and improved accuracy, now to within 0.25 mm. These new capabilities improved the workflow of the treatment procedure, dramatically shortening setup times as it was no longer necessary for the therapists to enter the room to adjust the patient setup. The capabilities did, however, require the development of entirely new QA procedures.

Geometric Quality Assurance of the Linac kV Imaging System

The addition of kV imaging to linacs introduced radiotherapy physicists to many of the QA test procedures used by diagnostic imaging physicists. In addition to the standard tests of image quality and imaging dose, there was also a safety consideration. The three arms that move the kV source, kV panel, and MV panel into position are physically large and easily capable of injuring a patient. Comprehensive checks on the proximity sensor and the more than 20 touch sensors arrayed around the imaging arms are performed daily. The most challenging task, however, was to develop QA tools for the geometric accuracy of the imaging system.

Image guidance, as implemented on the linacs, requires very accurate positioning of the imaging panels, as errors in panel positioning translate directly to errors in patient setup. In effect, the center of the panel must coincide with the projection of the radiation isocenter. The new TrueBeam linacs actively make small adjustments to the MV and kV panel positions as the gantry is rotated to correct for slight changes in sag that occur as the support arm changes angle. This creates a challenge for the QA physicist, as the standard tools available are less accurate than the system error being measured. To meet this challenge, we developed with the help of Ping Wang in Computer Services an automated QA procedure based on analysis of EPID images. It is essentially an extension of the traditional Winston-Lutz (WL) test of stereotactic setups, but now the MV WL images are linked to kV radiographs and cone-beam scans of the reference sphere. The displacement of the radiographic and CBCT imaging coordinate systems from the radiation isocenter can be determined to within 0.2 mm.

Clinical Impact of Image Guidance

Image guidance has had a profound effect on the practice of radiotherapy. Because of the clarity with which the patient anatomy can be visualized and compared with the reference image, setups are both more accurate and more clinically robust, greatly reducing the possibility of treating errors. With image-guidance, treatment margins could be reduced, allowing the physician to increase the biologically effective dose to the tumor while maintaining an acceptable normal tissue complication probability. Most importantly, the increase in geometric accuracy has led to new radiotherapeutic treatment strategies. Specifically, this has facilitated the implementation of SBRT (stereotactic body radiotherapy) or SABR (stereotactic ablative radiotherapy) in which fewer fractions (typically 1 to 5) are used, with much higher dose per fraction.

At MSKCC, Dr. Josh Yamada led the clinical introduction of the use of SBRT for the treatment of para-spinal metastases. The spine program has expanded greatly and now treats approximately 500 patients a year. Patients without prior radiation to the treatment site are treated to 24 Gy delivered in a single fraction; local control rates exceeding 90% are reported.

The SBRT program has grown to include treatments for other sites, including early-stage non-small-cell lung cancer, liver, pancreas, and prostate cancers. Over a quarter of all external beam treatment courses are now SBRT.

Brachytherapy—Further Development

In 1998, Dr. Marco Zaider was appointed Head of Brachytherapy Physics and Professor of Physics (in Radiology) at Cornell University Medical School. Under the leadership of Dr. Zaider, brachytherapy at MSKCC underwent a number of important changes. Manual treatment planning—in particular, for permanent prostate implants—was entirely replaced with intraoperative computer-based optimization of dose distribution. A second major change had to do with introducing—for the first time—*credible* biological (as opposed to dosimetric) plan optimization of brachytherapy treatment. This involves the notion that a radiation treatment plan should be evaluated in terms of its biological consequences—tumor control probability (TCP) and normal-tissue complication probability (NTCP)—rather than dosimetrically. Typically, a plan is evaluated in terms of a so-called *figure of merit* (FM) which quantifies the extent to which TCP is maximized while maintaining an acceptable NTCP.

As well, a number of novel treatment modalities (including associated QA procedures) were developed and introduced in the clinic. For instance:

1. Focal treatment of prostate cancer. Here the coordinates of the positive cores based on the mapping biopsy are recorded and correlated to the coordinates of the brachytherapy template. These coordinates are integrated into the intraoperative brachytherapy treatment planning system during the actual focal brachytherapy procedure to deliver the radiation prescription dose to the positive cancer regions.

2. Intraoperative radiation therapy for breast cancer using an MSKCC-developed applicator.

3. Intraoperative 32P high-dose-rate brachytherapy of the dura for recurrent primary and metastatic intracranial and spinal tumors; P-32 is imbedded in a solid foil.

4. Combined brachytherapy with external-beam radiotherapy.

5. Automated seed-finding algorithms (as opposed to manual identification of seed locations—a time-consuming and imprecise process).

6. An independent calculation program for the verification of high-dose-rate brachytherapy planning (QA).

In support of these projects, an emphasis was placed on developing biophysical models of radiation action geared toward radiation oncology. Thus:

1. A formulation of tumor control probability applicable to any temporal pattern of radiation delivery was developed.

2. The time evolution of normal-tissue complications was treated quantitatively, from the initiating event (injury) through eventual resolution. A system of differential equations describe the kinetics of injured cells (replacement and elimination), and whenever their number decreases under a lesion-specific threshold, the event is classified "complication."

3. The question of optimal fractionation schedule (same dose per fraction or changing dose?) was resolved by determining the optimal pattern of dose changes from fraction to fraction to result in optimal cure probability.

4. It was shown that the ubiquitous linear-quadratic description of cellular survival probability, which is the basis of all current treatment planning schemes, is in fact incorrect. A replacement was proposed.

Multi-Disciplinary Radiation Sciences Together Again

From the very beginning, following the discovery of x-rays by Röentgen and radium by Curie, the science of radiation had always been multidisciplinary. At Memorial, the team members of Failla, Quimby, and Laughlin included physicians, physicists, biologists, and chemists.

Eventually, however, the infrastructure at Memorial evolved into three independent clinical departments: radiotherapy, medical imaging, and medical physics. At other institutions, the multiple disciplines were usually under the umbrella of the department of radiology. However, in the 1960s and 1970s, as therapeutic radiology emerged as a discipline quite distinct from diagnostic radiology, independent radiation oncology departments were formed. In most institutions, medical physicists were also split along that divide, i.e., medical physicists were homed either in the department of radiology or radiation oncology.

However, medical physics remained as an independent department at Memorial Hospital. As described previously, in 1989 the department was organized into seven sections or services. Several sections of the department have center-wide responsibilities; other services or sections have assignments specific to individual clinical departments, primarily radiation oncology and radiology, but also for surgery and pediatrics. The intradepartment interactions between sections facilitate technology transfer to clinical departments, promote the application of a multiplicity of technical expertise to any given problem, and foster scientific excellence through mutual criticism. For the staff, the ability to participate in different applications of medical physics within one department provides stimulation and promotes a sense of shared mission.

New Role for Imaging in Treatment Planning

In the decades from 1980 to 1990, important advances in cancer imaging brought about somewhat of a re-integration of imaging and radiotherapy sciences. It was in that context that in September 1997 the National Cancer Institute organized an "Oncologic Imaging Workshop," convened by Carl Mansfield, M.D. (NCI Radiation Oncology), David Bragg, M.D. (NCI Diagnostic Imaging), and Robert Wittes, M.D. (NCI Medical Oncology).

The workshop's purpose was "to identify new opportunities for imaging in treatment planning…new research approaches and specifically find ways to enhance the rather limited practical clinical interactions between the imaging and radiation oncology communities when rich opportunities exist to tap the new advances in technology in both specialties."

Among the workshop's participants were physicists, CT/MRI diagnostic radiologists, radiation oncologists, and nuclear medicine physicians. Ling was invited to give a summation presentation and lead the general discussion to develop recommendations and action items for NCI to consider. Ling was initially reluctant to accept this responsibility as he felt inadequately informed in the diverse spectrum for discussion, and he knew that a good deal of effort would be necessary to become better prepared for the task. Although he conveyed his reservation to the NCI staff, the persistence and arm-twisting by his colleagues at the NCI persuaded him to accept the invitation. Here, fortunately, was a serendipitous opportunity to explore the horizon and then beyond.

Upon Ling's return from the NCI meeting, faculty members conducted workshops to consider the development of a research program focused on biological imaging for multi-dimensional conformal radiotherapy. These workshops culminated in a two-day workshop, April 9–10, 1998, in which Drs. David Bragg and Robert Sutherland, both Laughlin Visiting Professors, spoke on "Imaging in the 21st Century: A New Paradigm" and "Tumor Hypoxia and Gene Expression: Implications for Malignant Progression and Therapy." The Memorial faculty members who made presentations during the workshop were Koutcher, Kristin Zakian, Douglas Ballon, John Humm, and Ling.

Finally Together Again: Sciences of Radiological Imaging and Treatment

From that point on, momentum gathered for a multi-disciplinary biological imaging team effort. In 1999–2000, Ling obtained funding from the United States Army Medical Research to study "Non-invasive PET Imaging of Human Breast Xenografts: Influence of Tumor Hypoxia" with a direct cost of $221,000. With that fund, M. Urano, M.D., whom Ling knew from their MGH days, was recruited with the goal of developing animal models for biological imaging.

Further advances led to a five-year NCI RO1 Bioengineering Research Partnership grant (2001–2006) on "Multimodality Biological Imaging of Cancer and Tumor Hypoxia" with a direct cost of $2.6 M. Medical Physics faculty participating in that effort included Ling, Humm, Koutcher, Urano, Chui, Ronald Finn, Ph.D., Ballon, Kristen Zakian, Ph.D. Quihong He, Ph.D., Joseph O'Donohue, Ph.D., and Pat Zanzonico, Ph.D. As stated in that application,

> "The long-range goal of the study is to develop non-invasive multi-modality imaging that yields biological information of human cancers in a well-defined 3-dimensional coordinate system. The short-term objectives are to use nuclear magnetic resonance (NMR) and positron-emission tomography (PET) scanning to obtain multiple biological and 'hypoxia' images, which are spatially correlated due to a stereotaxic reference marker system, in rodent tumors and human xenografts. In addition, pO2 levels will be assessed with an oxygen probe in the same tumors, and tumor sections characterized to provide physiological and biological basis for understanding the NMR and PET images. This project integrates physical, chemical, biological, engineering and computer sciences to study tumor biology and hypoxia, with considerable significance for cancer diagnosis and treatment."

Finally, in 2006 an NCI five-year Program Project Grant on "Tumor Hypoxia Imaging—Laboratory and Clinical Studies" was awarded with a direct cost of $5.75 M. The leaders for the four research projects were Humm and Ling for Project 1, Koutcher for Project 2, Gloria Li for Project 3, and Jose Guillem (Department of Surgery, MSKCC) and Nancy Lee (Department of Radiation Oncology, MSKCC) for Project 4. Project 1 was on the "Detection of Tumor Hypoxia by Non-invasive Nuclear Imaging Methods" (see endnote 1). The title of Project 2 was "NMR Imaging of Tumor Hypoxia" (see endnote 2). The goal of Research Project 3 was to develop "Image-guided Gene-Radiation Therapy Targeting Hypoxic Tumors" (see endnote 3). Finally, Research Project 4 was a clinical study on "Clinical Imaging of Tumor Hypoxia in Rectal and Head/Neck Cancers" (see endnote 4). There were three Cores, led by Ling (Administrative), O'Donohue (Tumor Models), and Zanzonico (Animal Imaging).

The project eventually drew the attention of Dr. Harold Varmus, President of MSKCC (2000–2011) who included Imaging and Radiation Sciences as a program in the Cancer Center Core Grant, as described later.

The Harold Varmus and Robert Wittes Era

With the retirement of Dr. Paul Marks, the Board selected Dr. Harold Varmus as the President and CEO in January, 2000. Varmus, a Nobel prize winner as the co-discoverer of the human src protooncogene, was Director of the National Institutes of Health prior to his appointment at MSKCC. Upon his arrival, Varmus met with Clif Ling and asked him to explain the role of medical physicists. Their session was basically a discussion of the functions and activities of the department, and it included some discussion of Ling's own laboratory research on radiobiology and biological imaging.

As members of his management team, Varmus selected Dr. Robert Wittes as Physician-in-Chief of Memorial Hospital in 2001, and Dr. Thomas Kelly as Director of SKI. Wittes had previously been on the medical staff of Memorial hospital in the 1970s, and subsequently became a Deputy Chief at NCI/NIH, where he had been involved in grant funding. Wittes was aware of the medical physics research program at MSKCC and expressed his high regard for the department's work.

In preparation for the 2001 renewal of the Cancer Center Core Grant[*], Varmus asked various researchers with significant grant funding to make presentations so he could formulate strategy and construct scientific programs for application. He subsequently decided on two translational research themes, one of which was named Imaging and Radiation Sciences (IMRAS), to be led by Steve Larson, M.D., and Clif Ling. His decision was probably due the substantial extramural funding of various MSKCC faculty related to imaging and radiation sciences. Included in the program were investigators from Medical Physics, Radiation Oncology, and Radiology. Given the mandate and the perceived opportunity that it offered to strengthen radiological science research, Ling worked to construct a proposal, incorporating the input and ideas from the program members. At this point in time, the IMRAS was only a virtual program, without a formal structure. Its members were from various departments and were loosely connected. The Core Grant, including the Imaging and Radiation Sciences (IMRAS) Program, was favorably reviewed. The total funding of the Core

---

[*]MSKCC was the first institution to be designated as a *Comprehensive Cancer Center* under the National Cancer Act of 1971. As a Comprehensive Cancer Center, MSKCC was awarded a Cancer Center Core Grant from NCI/NIH, which is renewed periodically. The Core Grant supports the Core Facilities, such as Media Preparation and Flow Cytometry, which provides services common to many research programs.

Grant was increased significantly (some 30 to 40%) relative to previous periods.

In preparation for the 2003 budget, two reports were submitted to Dr. Wittes in October, 2002. The first was an analysis of faculty staffing and activity statistics from 1990–2003, and the second dealt with Research/ Development and funding sources. These reports indicated 19 faculty in 1990 (with two FTE funded by external research grants, i.e., soft money), and 30 faculty in 2003 (with 10 FTE funded by extramural research grants). The total faculty salaries in 2003 were $4.2 M, of which $1.4 M (or 33%) came from extramural sources. (The extramural research funding at this time did not include the Hypoxia Imaging Program Project Grant which began in 2006). Given that the department was functioning smoothly in both clinical support functions and research efforts, not much additional support was needed.

Dr. Wittes was in the process of moving from NCI to MSKCC and could be only a non-participatory observer during the Core Grant review. Upon the subsequent renewed funding of the Core Grant, Wittes encouraged Ling and his team to submit a proposal to Varmus for the formation of a formal IMRAS Program within SKI or MH with appropriate allocation of faculty, laboratory space, and institutional research funding. With input and suggestions from Wittes, the proposal was transmitted to Varmus in 2004. It was hoped that a formal IMRAS Program would be created, with allocation of laboratory space and operation budget, and staffed by an interdisciplinary team of full-time researchers of biologists, chemists, physicists, engineers, and clinical scientists.

In 2005, during the competitive renewal of the Core Grant, Ling discussed with Varmus the 2004 proposal and was told to go back and downscale the proposal, and suggested that the principals of IMRAS consider modeling IMRAS similar to the Experimental Therapeutics Center (ETC). This eventually resulted in the establishment of the IMRAS Program, which was basically an endowment of a sum of money used to fund intramural start-up projects.

Shortly after this time, during their monthly meeting, Ling indicated to Wittes his intention to step down from his position as Chair. A few hours later, Ling made this announcement in the Medical Physics senior staff meeting.

In July of 2007, Jean St. Germain, Attending Physicist and RSO, accepted the responsibility as Acting Chair, and Ling continued on a part-time basis as Primary Investigator of the Tumor Hypoxia Imaging Program Project funded by NCI/NIH. Upon the completion of that Program Project in the spring of 2012, Clif Ling became an emeritus member of MSKCC.

Endnotes

1. **Abstract of Research Project 1:** The overall goal of this project is to improve our understanding of the mechanisms and processes underlying tumor hypoxia images obtained with nuclear methods. Towards this goal, we shall use tumor models that contain a hypoxia inducible reporter gene to allow the cascade of molecular events induced by hypoxia to be either invasively or non-invasively visualized by optical and nuclear imaging methods. Specifically, the intra-tumor distribution of the expression of the reporter gene construct HRE-*tke*GFP: the fusion gene of the herpes simplex virus type 1 thymidine kinase *tk* and the enhanced green fluorescence protein *eGFP*, under the control of the hypoxia responsive element HRE will be imaged with microPET, digital autoradiography and fluorescence microscopy. Because the reporter gene is under the control of the HRE, we can determine the distribution of the initial hypoxia-induced event and provide a link between the biology of tumor hypoxia and other hypoxia-associated endpoints and surrogates. Specifically, the distributions of reporter gene expression will be compared with patterns of <u>endogenous</u> (e.g., HIF-1α and Ca9) and exogenous (e.g., pimonidzaole and EF5) hypoxia-associated markers, as assessed by immunohistochemical (IHC) analysis.

   In addition, microPET and autoradiography images of reporter gene expression will also be compared to the images of exogenous PET hypoxia imaging agents that are under clinical evaluation: ^{18}F-FMISO, ^{18}F-EF5 and ^{124}I-IAZGP. In linking hypoxia-induced molecular events to PET images, this study provides the first direct validation of non-invasive hypoxia imaging. As a reference, we shall use physical probes to perform intra-tumoral pO_2 measurements that are spatially co-registered with the images. We shall then apply these techniques to quantify the effect of interventions that modulate the effects of hypoxia including carbogen/hyperbaric oxygen breathing as well as re-oxygenation following radiation treatment.

   Finally, we propose to develop methods for <u>spatial co-registration</u> of the data from various sources: PET, autoradiography and NMR images, pO_2 probe data, tumor histology and IHC analysis of endogenous or exogenous markers, with the belief that the combined datasets from complementary methods will lead to an improved assessment of tumor hypoxia.

2. **Abstract of Research Project 2:** Treatment strategies are stratified by risk factors; a goal of stratification into different prognostic groups is to design risk based treatment strategies. Long term survival is determined by "risk factors" that predict prognostic parameters such as probability of local control and risk of developing systemic or metastatic disease. Tumor hypoxia is predictive of both risk of metastases and aggressive local disease. A non-invasive assay to measure hypoxia would provide prognostic information regarding local tumor aggressiveness and metastatic risk for development of risk based therapy, and for monitoring changes in oxygenation with treatment, could impact further therapy. The central hypothesis of this proposal is that tumor hypoxia and changes in oxygenation can be evaluated non-invasively using selected surrogate "markers". In all studies, we will utilize a stereotaxic template we have developed to register both the in vivo data and the p02 and pimonidazole studies. In Aim 1 we will test and validate a novel derivative of misonidazole (trifluoromisonidazole (T19F-FMISO)) for imaging hypoxia. We will determine the optimal dose as a balance between signal to noise requirements vs. specificity for imaging hypoxia, and also compare to 18F-misonidazole and pp02. In Aim 2 we will evaluate quantitation of lactate, dynamic contrast enhanced MRI, and T19F-FMISO as oxygen surrogates and validate them against p02 and pimonidazole. In Aim 2B, we will study changes in hypoxia induced by anti-neoplastic therapy and use these measurements as potential surrogate and compare to p02, microvessel density and radiobiological assays. The goal of Aim 2 is to determine which is the best surrogate of hypoxia and apply this in Aim 3. Aim 3 will use this data to optimize hypoxia driven suicide gene therapy. In aim 3, we will develop a fusion suicide gene (Cytosine Deaminase–Uracil Phosphoribosyl Transferase –CD-UPRT), under the control of a hypoxia response element (HRE) which can be quantitatively imaged by 19F NMR chemical shift imaging. We will develop this system as a both a reporter and suicide therapy system and evaluate its efficacy and compare it with thymidine kinase in parallel studies.

3. **Abstract of Research Project 3:** Tumor hypoxia, often present in human cancers, is associated with radio- & chemo-resistance, a more aggressive phenotype, and is prognostic of treatment outcome. The long term goal of this proposal is to develop image-guided radio-gene therapy, combining hypoxia imaging, radiation and gene therapy, to overcome hypoxia-mediated radioresistance and improve cancer cure. The major facets are 1) the imaging of tumor hypoxia, 2) image-guided delivery of vectors that express a radiosensitizing effector gene as well as a hypoxia-induced marker gene, 3) verification of delivery by molecular imaging, and 4) tumor eradication by radiation. To evaluate and validate this approach, there are three Specific Aims:

    In **Specific Aim I**, we shall design and construct vectors that express both a marker gene HSV1-*tk* from a hypoxia-inducible promoter and an effector gene that antagonizes the repair of DNA breaks, and use them to stably transfect tumor cell lines. The degree of radiosensitization of tumor cells that constitutively express various effector genes will be assessed *in vitro* and *in vivo*; the specific expression of the viral *tk* gene in hypoxic cells will be evaluated by microPET imaging of TK-mediated trapping of ^{124}I-FIAU in tumors transplanted in nude mice.

    In **Specific Aim II**, replication defective adenovirus containing the hypoxia marker gene and the effector gene will be constructed. The efficacy of adenovirus-mediated delivery of the marker gene and the effector gene, the hypoxia inducibility of the marker gene, and the radiosensitizing effect of the effector gene will be studied *in vitro* and *in vivo*. To improve virus dissemination and infection efficiency, we shall study conditionally replicative adenovirus and mild hyperthermia.

    In **Specific Aim III**, we shall transplant rodent and human tumors into nude rats, identify and localize the hypoxic sub-volume using microPET/^{18}F-FMISO imaging. Guided by these images, adenoviral vectors will be delivered to the hypoxic region of the tumor and microPET imaging based on TK-mediated trapping of ^{124}I-FIAU will be used to verify the preferential delivery of the vectors to hypoxic cells. The tumor's response to radiation will then be evaluated.

    In pilot studies, we have shown that antisense Ku70 or a dominant negative Ku70 fragment significantly radiosensitizes hypoxic tumor cells *in vitro* and *in vivo*, and that they can be successfully delivered into tumors by adenoviral vectors. For tumor hypoxia targeting, we have implemented microPET imaging with ^{18}F-FMISO and developed a stereotaxic template for guiding adenoviral vector injection. Also, we have shown that the hypoxia-induction of the marker *tk* gene can be readily detected using ^{124}I-FIAU / microPET *in vivo*. The information gained from the proposed preclinical studies will serve as a guide in the design of clinical strategy to improve the outcome of cancer therapy.

4. **Abstract of Research Project 4:** For rectal and head and neck cancers, direct pO2 measurements have established a significant intratumoral hypoxic fraction. This is of great clinical significance as tumor hypoxia is thought to negatively influence cancer response to chemoradiation (CMT), a major treatment component for these tumors. However, current methods for assessing tumor hypoxia are limited by their invasive nature and associated sampling errors due to heterogeneity of intratumoral oxygenation. The focus of this project is the clinical evaluation of non-invasive imaging of global tumor hypoxia via positron emission tomography (PET) in patients with rectal and head and neck cancers.

    We propose to evaluate two hypoxia tracers conjugated with positron-emitting radio-isotopes: 18F-FMISO and 124I-IAZGP. Given their different characteristics, difficult-to-predict pharmacokinetics, and possible disease and anatomy-related influences, we will test them in these two tumor types and identify the optimal tumor-specific tracer. Therefore, the Specific Aims of RP4 are:

    I. To identify the optimal hypoxia tracer for PET imaging of locally advanced rectal cancer and head and neck cancers. 100 patients of each disease site will be studied. Each patient will be imaged with 18F-FMISO and 124I-IAZGP. The images will be evaluated in terms of image quality and prognostic value of treatment outcome.

    II. To correlate, in rectal cancer, global tumor hypoxia assessed by PET imaging with that assessed by direct pO2 measurements using polarographic electrodes and IHC analysis of hypoxia-related proteins.

We believe that these series of clinical studies will demonstrate the utility of PET hypoxia imaging and enhance the management of patients with rectal and head and neck cancers.

References

Anderson, Lowell L. "Medical Physics and the Regional Medical Program." *Clinical Bulletin, MSKCC,* Vol. 1, No. 2, 1971.

Brucer, Marshall. "William Duane and the radium cow: An American contribution to an emerging atomic age." *Med Phys* 20:1601–05, 1993.

Chu, Florence, C. H. "A Personal Reflection on the History of Radiation Oncology at Memorial Sloan-Kettering Cancer Center." *Int J Radiat Oncol Biol Phys* 80(3):845–50, 2011.

del Regato, J. A. "Gioacchino Failla." *Int J Radiat Oncol Biol Phys* 19(6):1609–20, 1990.

del Regato, J. A. "James Ewing." *Int J Radiat Oncol Biol Phys* 2(1–2):185–98, 1977.

del Regato, J. A. "The Unfolding of American Radiotherapy." *Int J Radiat Oncol Biol Phys* 35(1):5–14, 1996.

del Regato, J. A. "Wilhelm Conrad Röntgen." *Int J Radiat Oncol Biol Phys* 1(1–2):133–39, 1975.

Hilaris, Basil S. "Brachytherapy: The first 100 years: An insider's views." Blurb, Inc., 2010.

Humm, J. L., et al. "In Memorium: John S. Laughlin, PhD (1918–2004)." *J Nucl Med* 46:26N, 2005.

Laughlin, John S. "Development of the Technology of Radiation Therapy." *Radiographics* 9(6):1245–66, 1989.

Laughlin, John S. "History of medical physics." *Physics Today* 36(7):26–35, 1983.

Laughlin, John S., J. Oradio, G. Shapiro, and Z. Abdun-Nabi. "High-Power Betatron for Cancer Therapy." *Electronics* Oct. 1953.

Quimby, Edith H. "Gioacchino Failla (1891–1961) and the Development of Radiation Biophysics." Twelfth Annual Meeting, Society of Nuclear Medicine. *J Nucl Med* 6:377–82, 1965.

Robison, Roger. "American radium engenders telecurie therapy during World War 1." *Med Phys* 27(6):1212–6, 2000.

St. Germain, J. and L. D. Rothenberg. "Obituary: John S. Laughlin." *Med Phys* 32:830, 2005.

Webster, Edward. "The origins of Medical Physics in the USA: William Duane, Ph.D., 1913–1920." *Med Phys* 20(6):1607–10, 1993.

APPENDIX: Biographies of Lead Authors, Editors, and Contributors

Lead Authors and Editors

Judith Groch

Surrounded by a family of physicians, Judith Groch has been a medical editor and writer for over 50 years. She is the author of the book *You and Your Brain*, a Harper and Row publication that won the 1963 Thomas Alva Edison Foundation Mass Media Award for the best science book for youth. She authored the highly acclaimed book *The Right to Create,* published by Little, Brown and Company in 1969.

In 1973, Judith joined the staff of the McGraw Hill publication *Medical World News* as a medical editor. When McGraw Hill discontinued the publication, she and many of her *Medical World News* colleagues joined the staff of *American Health Magazine*. After a decade or more at *American Health*, Judith moved to *Physician's Weekly*, which published a weekly medical news sheet for physicians' offices. She later joined the staff of *MedPage Today*, an online medical site, thereby transitioning to writing and editing web pages.

Judith then became a freelance medical writer and has had numerous pieces published in the Science Times section of *The New York Times*. She is the author of a children's book, *Play the Bach, dear*, published by Doubleday & Company. The project she holds most dear, however, is the history of medical physics at Memorial Sloan Kettering. Judith and her now deceased husband, Bill Minowitz, had formed close bonds and friendships with members of the Memorial Physics and Radiation Oncology Departments. Judith was honored to participate in the publication of this work.

John Laughlin

John Seth Laughlin graduated from Willamette University, received an M.S. from Haverford College, and got his Ph.D. in physics from the University of Illinois. While at Illinois he conducted research on both the betatron and the cyclotron. He continued this research at the University of Illinois College of Medicine as Assistant and then Associate Professor in the Department of Radiology. In Chicago his research focused primarily on the dosimetry of both photon and electron beams. It was here that he and colleagues made important contributions to the therapeutic use of high-energy x-rays and then pioneered the use of high-energy electrons.

John installed the first betatron dedicated to medical use at Memorial Sloan-Kettering (MSK), and in 1952 moved there as Attending Physicist and Chair of the Department of Medical Physics. He pioneered advances at MSK in treatment planning, dosimetry, and the early applications of computers in treatment planning. The resulting improvements in dose measurement allowed dose reductions for some forms of cancer. He also installed a cyclotron for the production of short-lived positron emitters used for both diagnosis and treatment.

John was a dedicated educator and established a training program for teaching hundreds of graduate students, residents, and radiological physicists. He was a Professor of Radiology at Cornell University Medical School from 1955 until his retirement. He served as Chief of the Sloan-Kettering Institute Department of Biophysics and was its Vice-President for a six-year term.

John served as President of a number of professional societies, including the AAPM, Radiation Research Society, Health Physics Society, and the International Organization of Medical Physics. He was a fellow of many other professional societies and served on the physics panel of the American Board of Radiology. He was Editor-in-Chief of *Medical Physics* (1988–1996) and AAPM historian (1979–2004). Other professional appointments include serving as a consultant to the Atomic Bomb Casualty Commission (1968–1974), the Los Alamos Laboratory (1968–1974), and the NYC Department of Health (1960–1978).

John was honored to receive many awards over the course of his career. These include the William D. Coolidge Award of the American Association of Physicists in Medicine (1964), the gold medal of the American Society for Therapeutic Radiology and Oncology (1993), the Distinguished Scientific Achievement Award of the Health Physics Society (1982), the Aebersold Award of the Society of Nuclear Medicine (1984), the gold medal of the American College of Radiology (1988), the Marvin M.D. Williams award of the American College of Medical Physics (1992), a distinguished alumni service award from the College of Engineering of the University of Illinois, and an honorary doctor of science degree from Willamette University (1964).

John's enthusiasms outside of physics included a dedication to his family and a love for competitive sports. The latter included squash, tennis, croquet, and the more cerebral chess, bridge, and general-knowledge word games. His more peaceful pursuits included sailing, a love of history, and an appreciation of classical and operatic music. He was a life-long member of the Society of Friends (Quakers) and always sought to see the inner light in all he met.

Jean St. Germain

Following completion of graduate study at Rutgers University and a fellowship at Brookhaven National Laboratory, Jean St. Germain was appointed in November 1967 as a fellow in the Department of Medical Physics at Memorial Sloan Kettering under John Laughlin and Garrett Holt. At the conclusion of her fellowship, she was appointed to the faculty and rose to the rank of associate attending physicist and, subsequently, attending physicist.

She served as the corporate radiation safety officer, guiding and presiding over the incredible growth of the institution. She served as interim chair of the Department of Medical Physics from 2007 to 2010 and subsequently as vice-chair for clinical and educational affairs and clinical member. Jean was a licensed medical physicist in New York State and was certified in comprehensive health physics in 1974 by the American Board of Health Physics (ABHP) and in 1991 in medical health physics by the American Board of Medical Physics (ABMP). Jean was also appointed a lecturer, instructor, and ultimately assistant professor of physics in Clinical Radiology, Weill College of Medicine, Cornell University and served as the radiation safety officer at the New York-Presbyterian Weill Cornell Medical Center for more than 35 years.

Jean's contributions to the field of medical physics and health physics were vast and significant. She has served several professional societies in key leadership roles. She served the American Association of Physicists in Medicine (AAPM) as national secretary, chair of the Rules Committee, parliamentarian, and founding chair of the Development Committee. She served four terms on the AAPM Board of Directors. She served as a member of the Governing Board of the American Institute of Physics, treasurer of the American Academy of Health Physics, chair of the Examining Panel in Medical Health Physics, and vice-chair of the American Board of Medical Physics. In the Greater New York area, she served as president of the Radiological and Medical Physics Society (RAMPS, the New York City chapter of the AAPM) and served three terms as president of the Greater New York Chapter of the Health Physics Society (HPS). Jean was a member of the scientific committee (SC) for the National Council on Radiation Protection and Measurements (NCRP) that produced NCRP Report No. 105 on radiation protection of medical and allied health personnel. She later served as chairman of the SC that produced NCRP Report No. 155 on the management of radionuclide therapy patients. In addition, Jean served as a member of several New York State advisory committees on medical and radiological health. She also served as a special examiner for the New York State Civil Service Commission.

Jean received many honors and awards during her career. She was a fellow of the HPS and of AAPM. She was presented the Failla Award by the Greater New York Chapter HPS and RAMPS. She received the AAPM Distinguished Service Award in 2001, as well as the Varian Award for best professional paper in the *Journal of Applied Clinical Medical Physics* in 2004. And in 2015, Jean was presented with the Marvin M.D. Williams Professional Achievement Award by the AAPM. The award recognizes AAPM members for an eminent career in medical physics with an emphasis on clinical medical physics.

Jean was an excellent lecturer and teacher. She taught health physics and radiation safety to generations of medical physicists, radiologists, radiation oncologists, nuclear medicine physicians, radiotherapists, radiologic technologists, lab scientists, and others at Memorial Sloan Kettering Cancer Center, Weill Cornell Medical Center, and throughout the medical physics and radiological community.

Beyond physics, Jean's great passion was music. She took vocal lessons at Julliard and was an operatic soprano soloist who gave many recitals and concerts throughout her life. She was also a regular attendee of performances at the Metropolitan Opera. Her commitment to service extended to her church, St. Joseph's in Yorkville, where she was a trustee. She was also an active member of the National Society of Arts and Letters, serving on the Winston Scholarship Committee and the Shirley Rabb Winston Scholarships in Voice.

Jean considered Marie Curie a heroine, as she was the first woman to win a Nobel Prize and then became the first person to attain a second Nobel Prize. Jean had the pleasure of spending time with Marie Curie's daughter, Eve Curie Labouisse, who had written an extensive biography of her mother.

C. Clifton Ling

C. Clifton Ling was born in Quilin, China and received his primary and secondary education in Hong Kong at the Tak Sun School and La Salle College, respectively. He attended universities in the United States, receiving his B. Sc. (magna cum laude) at Oregon State University and his Ph.D. in nuclear physics from the University of Washington in 1971.

He then entered radiation biophysics as a Research Fellow at Memorial Sloan-Kettering Cancer Center. Since then, he has held academic appointments at the Massachusetts General Hospital and Harvard Medical School, George Washington University Medical Center, and the University of California, San Francisco. In 1989, he returned to MSKCC as the Enid A. Haupt Professor and Chairman of the Department of Medical Physics and Professor of Radiology (Physics), Weill Cornell Medical College of Cornell University. At Memorial Sloan Kettering, he succeeded the late Gioacchino Failla and John Laughlin as Chair of the Medical Physics Department. In 2007 Dr. Ling joined Varian Medical Systems as Director, Advanced Clinical Research, while maintaining his research at MSKCC.

Dr. Ling has been an active participant in many professional societies, such as the AAPM, the Radiation Research Society, and ASTRO. In AAPM, he served on the Board of Directors (1982–1987) and chaired the Scientific Program Committee (1983–1987) and the Science Council (1991–1993). He chaired the ASTRO Radiation Physics Committee, and was a councilor in physics in the Radiation Research Society. He was on grant review panels of both the United States and Canadian National Cancer Institutes, and was on the Nuclear and Radiation Studies Board of the National Academies. He has also been on the editorial boards of *Medical Physics*, the *International Journal of Radiation Oncology, Biology, and Physics*; *Radiotherapy Oncology*; *Seminars in Radiation Oncology*; and *Radiation Research*.

Dr. Ling has received numerous honors and awards, including Honorary Member of ESTRO, the Ray Bush Visiting Professor of Princess Margaret Hospital, the Suntharalingam Lecturer of Thomas Jefferson University, Speaker of the Royal College of Physicians and Surgeons of Canada, the Ira Spiro Visiting Professor of Harvard Medical School, the Franz Buschke Lecturer of the UCSF, and the James Purdy Lecturer of the Washington University, St. Louis. He received the AAPM Coolidge Award in 2004, the Gold Medal from ASTRO in 2006, and the Lifetime Achievement and Contribution Award from the Radiation Oncology Society, Republic of China, 2007. In 2012, he was the Failla Memorial Lecturer of the Radiological and Medical Physics Society (RAMPS).

Dr. Ling's research interests range from the fundamentals of cancer radiation biology to optimized radiation treatment planning and delivery, and more recently biological imaging as applied to cancer management. He has contributed to brachytherapy dosimetry, particularly of ^{125}I seeds. In collaboration with other scientists and clinicians, he has participated in the development of 3D-CRT and IMRT, helping usher in the widespread use of these advanced techniques. In biological research, Dr. Ling has studied oxygen effect, dose rate effects and the repair of sublethal damage, hypoxic cell radiosensitization, radiation-induced carcinogenesis and apoptosis, and the effects of oncogenes on radiosensitivity. Subsequently, his laboratory focused on the biological basis of molecular and functional imaging.

Dr. Ling has authored nearly 290 peer-reviewed papers and over 30 chapters in books and proceedings. He has been the principal investigator on numerous grants from the National Institutes of Health, the Department of Energy, the Department of Defense, and the American Cancer Society

Additional material of C. Clifton Ling in historical interviews by AAPM and ASTRO can be viewed in the following links:

https://www.aapm.org/org/history/InterviewVideo.asp?i=93.
https://www.astro.org/About-ASTRO/History/C-Clifton-Ling.

Contributors

Lowell Anderson

Lowell Anderson received his Ph.D. degree in biophysics from the University of Rochester in 1958. After an 11-year appointment as biophysicist at Argonne National Laboratory, where he worked mainly on developing instrumentation for neutron dosimetry, he joined the Department of Medical Physics at Memorial Sloan-Kettering Cancer Center (MSKCC) in New York. At Memorial, because of his neutron dosimetry background, he was selected to coordinate a contract with the Department of Energy to evaluate the use of ^{252}Cf neutron sources in interstitial brachytherapy. That project led to his long-time interest and specialization in brachytherapy physics. Lowell was head of Brachytherapy Physics until his retirement in 1998. He is currently Member, Emeritus at MSKCC.

For nearly 20 years, Lowell served as the coordinator of a lecture course in medical radiation physics at MSKCC, and as a lecturer in basic radiation physics and in brachytherapy physics. Subsequently, he continued to give the lectures in brachytherapy physics. Participants in the course included MSK residents in radiation oncology and fellows in medical physics, as well as residents in radiology from Cornell University Medical College, where Lowell held an academic appointment. He also supervised a number of degree candidates and post-doctoral fellows in various brachytherapy physics research projects.

Lowell conducted an active program of research and development in brachytherapy physics. In the days before 3D imaging was available on which to base implant planning, he constructed nomographs to guide permanent implant procedures using ^{125}I and ^{103}Pd seeds and temporary implants of ^{192}Ir seeds in ribbons. He performed or collaborated in dose distribution measurements for ^{137}Cs, ^{192}Ir, ^{125}I, ^{103}Pd, and ^{252}Cf sources. He developed least-squares optimization techniques for high-dose-rate (HDR) remote afterloading. He published a planning system for stereotactic temporary implants of brain tumors with high-strength ^{125}I seeds involving least-squares optimization of seed positions. He applied a variation of this system to surgical-deficit mold treatments. He conceived the idea for, and wrote the software to implement, a "natural" ($-3/2$ power) dose-volume histogram to assist implant design and evaluation. Finally, he directed the compilation of an extensive treatment-planning atlas for intraoperative HDR remote after-loading using quasi-planar, flexible-plastic applicators. He authored or co-authored 80 papers in peer-reviewed journals, as well as two books and 52 book chapters.

Chandra Burman

Chandra Burman received his Ph.D. in Physics from the State University of New York at Albany in 1980. He worked there as a Research Associate in experimental physics to study solid surfaces using ion beams. One of his interests was the study of the impact of radiation damage on the durability of glass. In 1984, he joined the Memorial Sloan Kettering Cancer Center as a postdoctoral fellow. After completion of the fellowship in 1986, he worked as a Senior Scientist on the NCI- sponsored Photon Project: "Evaluation of High Energy Photon External Beam Treatment Planning." Under the program project, state-of-the-art three-dimensional photon treatment planning was developed. In collaboration with John Lyman and Gerald Kutcher, he developed the LKB normal tissue complication model. This model predicts the normal tissue complication probability (NTCP) for non-uniform irradiation. He was instrumental in the development and implementation of the 3D treatment planning system for clinical use at MSKCC. In 1987, headed the 3D treatment planning group. Later, he worked on the development and clinical implementation of intensity-modulated radiation therapy (IMRT). The first IMRT patient was treated at MSKCC in 1995. Chandra Burman has presented his work at international meetings and given refresher courses and workshops on 3D and IMRT planning.

In 1995, MSKCC established regional cancer care centers to offer cancer care at convenient locations surrounding the main campus in Manhattan. In 1997, he was appointed director of the regional physics section. His responsibilities included supervision of the physics staff at all the regional sites. In 2003, he was promoted to the rank of Attending Physicist.

He has been active in the local AAPM chapter, Radiological Medical Physics Society (RAMPS) of New York. He served as the president of RAMPS in 2009. He has received the Distinguished Medical Physicist award from the Indo-American Society of Medical Physicists, and he is a Fellow of AAPM. He has authored over 90 peer-reviewed papers/book chapters.

Jenghwa Chang

Jenghwa Chang earned his Ph.D. degree in Electrical Engineering from the Polytechnic University of New York in 1995. He was a Research Assistant Professor of Pathology at SUNY Health Science Center at Brooklyn from 1993–1997, an Assistant/Associate Attending Physicist of Medical Physics at MSKCC from 1997–2008, an Associate Professor and the Director of Physics Research of Radiation Oncology at NYU Langone Medical Center from 2008–2010, and an Associate Professor and the Director of Centralized Treatment Planning of Radiation Oncology at New York Presbyterian Hospital from 2010–2016. Dr. Chang is currently an Associate Professor and a Senior Medical Physicist at the Department of Radiation Medicine of Northwell Health and Donald and Barbara Zucker School of Medicine at Hofstra/Northwell.

Dr. Chang was certified by the American Board of Radiology in 1998 and has been practicing medical physics for over 20 years. Dr. Chang was a member of the AAPM Therapy Physics Committee from 2006–2008 and served as the president of RAMPS in 2011. He is a site surveyor for the ACR Radiation Oncology Practice Program and a reviewer for multiple international journals. Dr. Chang has been involved in the training of technologists, medical residents, and physics residents since 2000. He was the co-founder of the medical physics residency program at New York Presbyterian Hospital and served as the Program Director at the Weill Cornell Medical College campus from 2010–2016. He is currently the Director of Medical Physics Residency Program of Northwell Health and the coordinator of radiation oncology physics course for medical residents there.

Dr. Chang's research interest involves applying engineering and physics principles to medicine, particularly in the fields of radiology and radiation oncology. At SUNY HSCB, Dr. Chang was a pioneer in optical diffusion tomography for early detection of breast cancers. During his tenure at MSKCC, he investigated the electronic portal imaging device for verification of IMRT fields, implemented an in-house mega-voltage cone-beam computed tomography on a medical linear accelerator to improve the treatment setup accuracy and critical organ avoidance for radiation oncology patients, and developed algorithms for registering MRSI and functional MRI for radiotherapy of gliomas. His research at NYPH/WCMC involved the development of a low-cost radiation delivery unit, as well as the optimization of treatment planning and delivery processes for radiotherapy. Recently Dr. Chang became interested in the application of automation and artificial intelligence to treatment delivery and planning to improve the quality and efficiency of radiotherapy. His research efforts in the past three decades have produced 59 peer-reviewed papers, 45 book chapters/proceeding papers, more than 120 abstracts in conferences of major professional societies, and two patent disclosures. Dr. Chang is a member of IEEE, AAPM, and ASTRO.

Chen-Shou Chui

Chen-Shou Chui received his Ph.D. degree in Applied Physics and Nuclear Engineering from Columbia University in 1985. He joined Memorial Sloan-Kettering Cancer Center in 1979 as a senior programmer in the Computer Service of the Medical Physics Department and gradually rose to the rank of attending physicist and member of Memorial Sloan-Kettering Cancer Center.

His work focused on the applications of computers in radiation therapy and radiation dosimetry. Collaborating with Radhe Mohan, he developed three-dimensional CT-based treatment planning systems. They used Monte Carlo methods to study the physical characteristics of clinical photon beams, which served as the basis for more advanced dose calculation methods, such as pencil beam convolution and differential pencil beam for tissue inhomogeneity correction. Chen-Shou also applied Monte Carlo methods to dosimetry problems in nuclear medicine and radiation therapy.

From 1989 to 1994, Chen-Shou was appointed the head of Radiation Dosimetry, where he was also in charge of the Regional Dosimetry Laboratory. During this period he participated in the NCI-sponsored program project "3D Conformal Radiation Therapy" led by Clifton Ling. Collaborating with Tom LoSasso, Chen-Shou worked on problems related to the dosimetry of multileaf collimator.

Chen-Shou was appointed the Chief of Computer Service in 1995. Collaborating with his graduate student Spiridon Spirou, he worked on a new treatment technique—intensity-modulated radiation therapy (IMRT)—and successfully implemented it in the clinic. The optimization algorithm of IMRT was based on the conjugate gradient method, which took into account dose-volume constraints. The optimized intensity distribution was delivered using the dynamic motion of the multileaf collimator. This work culminated in the world's first delivery of IMRT using a multileaf collimator at MSKCC on September 17, 1995.

Chen-Shou had served on various task groups and the Radiation Therapy Committee of the American Association of Physicists in Medicine (AAPM), and was elected a fellow of AAPM in 2000. He served as associate editor of the journal *Medical Physics* from 1995 to 1998. Chen-Shou had supervised a number of graduate students and post-doctoral fellows, and he has lectured on courses for the residents at MSKCC and graduate students at Columbia University. He had published more than 100 papers in peer-reviewed journals and written 15 book chapters. He received Farrington Daniels awards of the *Medical Physics* journal in 1987 and 1989, and awards for excellence of the *Journal of Applied Clinical Medical Physics* in 2002 and 2004.

Chen-Shou was board certified by the American Board of Radiology in Therapeutic Radiological Physics and licensed in New York State for therapeutic radiological physics.

Lawrence T. Dauer

Lawrence Dauer initially earned a B.A. in Biology and Chemistry in May 1987. During his undergraduate studies, he interned for two years at the Indian Point Nuclear Power Plant and joined the Greater New York Chapter of the HPS, where he initially worked with Jean St. Germain of MSKCC. Upon graduation, he served as a health physicist/radiochemist at CintiChem, Inc., a facility producing radioactive materials for medical diagnostic use. He returned to Indian Point in radiological engineering roles from 1990 until 1999, eventually serving as Supervisor. He went on to receive his M.S. in Health Physics from the Georgia Institute of Technology, Atlanta in March 1996 and was certified in comprehensive health physics by the American Board of Health Physics in 1999. He was later awarded his Ph.D. in Adult Education from Capella University, Minneapolis in October 2005. After co-owning a radiological engineering consulting company, he joined the Radiation Safety Section of MSKCC as Radiation Safety Operations Manager in September 2002 at the urging of Jean St. Germain. He was appointed to the faculty as Assistant Attending Health Physicist in 2007 and Associate Attending Physicist in 2013 with a primary appointment in the Department of Medical Physics and a secondary appointment in the Department of Radiology.

Larry concentrated his research activities on radiation protection in the fields of radiology, x-ray imaging, nuclear medicine, and radiation oncology, with emphasis on organ doses and epidemiology. Larry served as Chair and Vice Chair of the Radiation Safety Committee of the AAPM, President and Executive Council Member of the Medical Physics Section of the HPS, President of the Greater New York Chapter of the HPS, and Board Member of the Radiological and Medical Physics Society of New York. He is currently a Council and Board member of the National Council on Radiation Protection and Measurements (NCRP). He served seven years on the International Commission on Radiological Protection (ICRP) Committee 3– Radiation Protection in Medicine, and is a member of the Science Committee of the International Organization for Medical Physics. He was a member of the Institute of Medicine/National Academies Committee on Research Directions in Human Biological Effects of Low Level Ionizing Radiation and has served as a consultant to the International Atomic Energy Agency. Larry received both the Elda Anderson Award and a Fellow Award from the Health Physics Society. In 2018, Larry was named the Corporate Radiation

Safety Officer following the untimely death of Jean St. Germain, leading a team of extraordinary Health Physicists providing clinical, research, and regulatory support to the institution.

Larry remains active in both ICRP and NCRP, producing several radiation protection guidance documents, including the following: co-Chair of NCRP Commentary No. 26—Guidance on Radiation Dose Limits for the Lens of the Eye; co-Chair of NCRP Commentary No. 27—LNT and Radiation Protection; member of ICRP Publication No. 98—Radiation Safety Aspects of Brachytherapy for Prostate Cancer Using Permanently Implanted Sources; member of ICRP Publication No. 139—Occupational Radiological Protection in Interventional Procedures; and co-Chair of ICRP Task Group 89 on Radiological Protection for Staff in Brachytherapy.

Edward Epp

Edward Epp was born in Saskatoon, Saskatchewan in Canada on July 21, 1929. He obtained his M.A. at the University of Saskatchewan in 1952. As a graduate student he worked with Professor H. E. Johns, then the leading medical radiation physicist in Canada. In those days, commercially made instruments were relatively rare. So the Johns' Lab made its own such as ionization chambers and DC amplifiers for making depth dose measurements.

In 1952, Epp worked at the National Research council of Canada in Ottawa to gain further experience in instrumentation. In 1953, he began his studies of nuclear physics in Montreal, obtaining his Ph.D. in 1955 at McGill University under Professor John S. Foster, Head of the Physics Department, using the 100 MeV cyclotron. Epp worked on the decay schemes of rhenium and osmium using beta and gamma ray spectrometers. In 1955, Epp was offered a position in the Radiology Department of the Montreal General Hospital where he would supervise the installation of the new ^{60}Co unit for the treatment of cancer. ^{60}Co was the latest word in radiation treatment at that time. Epp had gained experience in ^{60}Co in the laboratory of H. E. Johns, who was one of the inventors of ^{60}Co for cancer therapy. In 1957, Epp was offered a position at Memorial Sloan-Kettering Center in New York to do research on various topics suggested by Professor John S. Laughlin, which included the measurement of the spectral photon distribution of diagnostic x-rays. At this time Epp became interested in working on the oxygen effect in cancer treatment. Using 0.5 MeV electrons from field emission sources, a single pulse could be delivered in as short a time as 3 nanoseconds at ultra-high dose rates. Double doses delivered with time separations over a wide range of microseconds delivered to thin layers of bacterial and mammalian cells were found useful in studies of the mechanism of the oxygen effect.

In 1975, Epp became a Full Professor at Harvard Medical School. He was also appointed Head of Radiation Physics in the Department of Radiation Oncology at Massachusetts General Hospital. At MGH Epp was in charge of modernizing the radiation delivery aspects to a fully state-of-the-art facility using linear accelerators as the primary radiation delivery systems. Dr. Epp is grateful to have received unbroken funding over a three-decade period by the National Cancer Institute.

Dr. Epp retired in 1998 and lives with his wife, Shirley Mae, in Needham, Massachusetts.

Doracy P. Fontenla

Doracy P. Fontenla did her physics studies at the University of Rio de Janeiro, Brazil. After completing her B.S. degree in physics she received an UNESCO fellowship to do her post-graduate studies at the Instituto Balseiro, in Bariloche, Argentina, where she was a pioneer in the field as one of only three women in the program, and the only woman to graduate from the program. After graduation from the Instituto Balseiro, she obtained a fellowship at the University of Florida in Gainesville, Florida, where she was a doctoral student. During this period she worked on the research for her doctoral thesis in nuclear reactions. She submitted her thesis work at the Instituto Balseiro where she was awarded her Ph.D. degree in Nuclear Physics in 1968, the only female to graduate in her class as a Nuclear Physicist. She then worked for several years at the Instituto Balseiro in Argentina, at the Physics Department of the University of Sao Paulo, and at the University of Campinas in Brazil, starting as an Assistant Professor and reaching the position of Associate Professor in the Physics Department of the University of Sao Paulo and the University of Campinas, in Sao Paulo, Brazil.

While on a leave of absence from the University of Campinas, Dr. Fontenla made several visits to the Brookhaven National Lab in Upton, NY as a guest research scientist and was asked to join the Brookhaven National Laboratory nuclear physics research group, where she worked from 1977–1979. In 1980, she joined the Department of Medical Physics at Memorial Sloan Kettering Cancer Center (MSKCC) in New York as a postdoctoral fellow. After graduation from the post-doctoral program, she was invited to join the staff of the Medical Physics Department at MSKCC, where she was given responsibility for various positions: Director of the Memorial Accredited Dosimeter Calibration Laboratory (1984–1991), Assistant Prof. in Physics of Radiology at Cornell University Medical College (1987–1993), Assistant Attending Physicist (1988–1999), and Chief of External Beam Treatment

Design MSKCC (1990–1993). During this period at MSKCC, among other responsibilities, Dr. Fontenla worked on the improvements of various clinical techniques, such as TBI, TSEB, and *in vivo* dosimetry for patients. In 1993, at the invitation of Dr. Badrasaim Vikram, who was leaving MSKCC's Radiation Oncology Department to become the Director of the Radiation Oncology Department of Montefiore Medical Center, Dr. Fontenla accepted the position of Associate Professor and Director of Medical Physics at the Albert Einstein College of Medicine. In 2000, coinciding with Dr. Vikram's departure from Montefiore to join the IAEA, Dr. Fontenla was invited to join the medical physics section at Long Island Jewish Medical Center as the Associate Medical Physics Head, where she worked until 2008.

In 2009, when the Medical Physics Department at MSKCC was looking to apply for CAMPEP Accreditation of their Medical Physics Residency Program, Jean St. Germain, at that time the Acting Chair of the MSKCC Medical Physics Department, invited Fontenla to re-join MSKCC to direct that initiative. She accepted the position as the Director of Medical Physics Residency Program at MSKCC. Under her leadership the CAMPEP application was approved in 2009. The program was successful and later reapproved in 2015.

Aside from her work at various major hospitals in the United States, Dr. Fontenla is very well known and respected internationally, particularly in Latin America where, as AAPM Chair of the Latin American Affairs Committee and ISEP, she organized several symposia and medical physics meetings. Additionally, in collaboration with IAEA, she participated as speaker at Argonne National Laboratory, Argonne, IL, in several Regional IAEA Training Courses on different aspects of Radiation Therapy for Radiation Therapy Technologists, Medical Physicists, and Medical Physics Residents.

Dr. Fontenla also participated, as a member of the New York Licensure Committee, in the effort to obtain Medical Physics Licensure in the State of New York. He continues to have an active participation in the field, contributing in various AAPM committees and being invited to give talks internationally. Finally, in addition to her professional accomplishment, Dr. Fontenla raised five children, three of whom either were or are affiliated with the field of medical physics.

John Humm

John Humm received his Ph.D. in biophysics from the Polytechnic of the South Bank, London in 1983 for a thesis on the microdosimetry of Auger electron-emitting radionuclides conducted at the Kernforschungsanlage in Jülich, Germany. His first job was at the Medical Research Council in Harwell, Oxfordshire, England where he worked on the dosimetry of tumor-targeting monoclonal antibodies. In 1988, he spent a one-year term in the Department of Medical Oncology at Charing Cross Hospital to assist with the dosimetry of the new radiolabeled monoclonal antibody program.

In 1989, he came to the United States, where he was recruited to Harvard Medical School by radiation oncologist Roger Macklis and medical physicist Lee Chin to assist building the radioimmunotherapy program there for the treatment of lymphoma. He also played a key role in support of the radiobiology program of Dr. Macklis to explore the potential of α-emitting radioimmunoconjugates. During this time, John also learned the duties of a radiotherapy physicist, for which he would later become ABR certified.

In 1993, Clif Ling recruited Humm to join MSKCC. Initially he was placed in the dosimetry group under Chen Chui, where he developed software for collision avoidance and assisted in the quality assurance of the MM50 racetrack microtron. In August 1995, Clif promoted John to lead the Nuclear Medicine Physics group. John conducted an active program of clinical support, as well as an active research effort, and had to sit the boards again, this time in Nuclear Medical Physics. In those days Nuclear Medicine consisted of only four attending under the leadership of Dr. Steven Larson, and John's group consisted of George Sgouros, Farhad Daghighian, and Hovanes Kalaigian. John hired a post doc, Yusuf Erdi (who later took over as chief of the diagnostic physics group from Larry Rothenberg) and inherited a graduate student, Osama Mawlawi, who because the chief of Nuclear Imaging Physics at MD Anderson. The great concordance of interest in targeted radionuclides from Dr. Larson, Dr. Lloyd Old (who led the Ludwig Institute of Cancer Research), David Scheinberg (leukemia service), Nai Kong Cheung (pediatric oncologist), and others was the platform on which John, George Sgouros, and Joe O'Donoghue built what was to become perhaps the leading radionuclide dosimetry physics group in the world.

In 1995, MSKCC purchased its first whole-body PET scanner. John and Yusuf Erdi, Sadek Nehmeh, and Brad Beattie built a strong PET physics program. One of the crowning achievements of this period was the development of respiratory-gated PET, which was pioneered at MSKCC. Interest in the use of PET for radiotherapy planning led to significant advances in quantitative hypoxia imaging by PET.

John Humm has published over 200 peer-reviewed articles on a wide range of topics. His greatest contributions have been in Auger and alpha particle dosimetry, optimization of targeted radionuclide therapies, as an early pioneer in digital autoradiography, and for his work in PET imaging— in particular on the quantitative determination of the spatial distribution of tumor hypoxia by fluoromisonidazole PET imaging, a methodology that has shown promise in new dose de-escalation protocols in H&N cancers led by radiation oncologist Nancy Lee.

He has contributed to numerous NCI grants. He was awarded an R01 for optimizing the combination of radioimmunotherapy with external beam radiotherapy and was a project leader in a P01 grant by Dr. Ling entitled "Tumor Hypoxia Imaging: Laboratory and Clinical Studies." Most recently, he was awarded two R01 grants on (1) preclinical imaging drug delivery during chemoradiation therapy of pancreatic cancer and in a multi-leadership R01 with Steve Larson and Mike Tuttle (2) "^{124}I-NaI PET: Building block for precision medicine in metastatic thyroid cancer."

Andrew Jackson

Andrew Jackson obtained his Ph.D. in theoretical particle physics from London University in 1984, studying the interaction between nucleons as modeled by that between Skyrmions. Between 1984–1990 he pursued these ideas during post-doctoral fellowships in theoretical nuclear physics at Stony Brook University and the Lawrence Berkeley Laboratory.

In 1990, he joined the Medical Physics Department at Memorial Sloan Kettering Cancer Center as a post-doctoral fellow and studied dose calculation algorithms, treatment planning optimization, and modeling treatment outcomes with Radhe Mohan, Gerry Kutcher, Ellen Yorke, and Clifton Ling. From 1990–1993, he studied outcomes of inhomogeneous irradiation of parallel organs. He became instructor in Medical Physics in 1992. In 1995, he was the first to analyze the incidence of complication probability in patients treated with radiotherapy for liver cancer using patient-specific outcomes and dose-volume histograms. In 1995, he was appointed Clinical Assistant in Medical Physics. From 1990 to 1994, he wrote the plan evaluation module of MSKCC's new treatment planning system, created by Radhe Mohan and Chen Chui, in use from 1994–2014. In 1995, in studies with Radhe Mohan and Xiao-Hong Wang, he used insights gained from studying volume effects in normal tissue tolerances to establish the need for dose volume constraints in automated optimization of treatment planning for IMRT. From 1997 to 2007, he was co-project leader, with Dr. Steven Leibel, of the Clinical Studies project of Clif Ling's interdepartmental P01 grant (Intensity-Modulated Radiation Therapy), responsible for research project 1b: Dose Distributions and Outcomes Analysis. In 1998, he was appointed Assistant Attending Physicist. From 1995–2007, in collaboration with physicians from the department of Radiation Oncology and physicists from the treatment planning service of the Medical Physics department, he applied the statistical methods he pioneered to analyze dose-volume and outcome data from 3DRT and IMRT treatments for prostate, head and neck, and NSCLC, and he applied the tolerances derived to the dose escalation trials conducted under Clif Ling's P01 grant at MSKCC. He initiated the MSKCC planning system's transfer to a PC-based platform when he was head of the treatment planning group of Computer Services from 2001–2002. In 2003, he was appointed Associate Attending Physicist.

From 2007 to 2010, Andrew served on the steering committee of QUANTEC, reviewing and synthesizing tolerance doses for treatment planning of radiotherapy. From 2009–2015, he was PI of an R01 grant to use advanced reporting methods to encourage meta-analysis of dose-volume dependence of radiotherapy outcomes. From 2010–2015, he was consortium leader for an R01 grant on the rapid and efficient constrained optimization of IMRT and VMAT Treatment Planning, collaborating with Prof. Rich Radke of Rensselaer Polytechnic Institute. In 2010, he became chair of AAPM's Biological Effects Sub-Committee (a position he holds to this day). From 2010–2012, he was instrumental in founding the HyTEC and PENTEC international collaborations; he currently serves of the steering committees of both these efforts, applying the QUANTEC methodology to hypofractionated and pediatric treatments. As part of his ongoing studies of the influence of dose-distributions on the outcomes of radiotherapy, he is currently investigating the influence of thoracic irradiation on the immune system and consequent effects of overall survival, local regional failure, and progression free survival in non-small-cell lung cancer and esophageal cancer patients.

In the course of his career in medical physics, Andrew has mentored and supervised 12 doctoral students, post-doctoral fellows, and research assistants, including: Mark Skwarchuck, Sabine Levegruen, Joseph Bauer, Reshma Munbodh, Renzi Lu, Fan Liu, Hans Stabenau, Eric Williams, Ming Yan, Linda Rivera, Shiama Bakr, and Maria Thor.

Gloria C. Li

Dr. Gloria C. Li was born in China and grew up in Taipei, Taiwan. After receiving her B.Sc. degree in physics from the National Taiwan University, she traversed the Pacific to continue her graduate studies, earning a doctorate in high- energy nuclear physics from Stanford University in 1972. Between 1972 and 1980, she remained at Stanford to carry out research on cancer treatment by pi-mesons (pion), as well as research on hyperthermia cancer therapy. She designed and implemented a 3D treatment planning system for focusing the negatively charged pion particles from the 60-channel Stanford Medical Pion Generator into the tumor. Her hyperthermia work dealt mainly with a phenomenon termed thermotolerance, by which cells become resistant to higher temperatures, and thus acquire impunity to hyperthermia therapy. Her research led to the development of a model that accounts for the induction, development, and decay of thermotolerance in cells.

In 1981, she was recruited to the University of California, San Francisco. She continued her research on cellular responses to heat-shock and radiation, and rapidly rose to the rank of full professor. She pioneered functional studies of the mammalian heat-shock protein 70, and elucidated the mechanistic basis of thermotolerance. In 1990, she joined the Memorial Sloan-Kettering Cancer Center (MSKCC) as Head of the Laboratory of Radiation and Hyperthermia Biology, was appointed Member with Tenure of Title at MSKCC, and concurrently served as Professor of Physiology and Biophysics at the Weill Cornell Medical College. There, while unraveling the mechanism of the heat-shock response in cells, she discovered the involvement of a dimeric protein Ku70/Ku80 in heat-shock protein regulation. The Ku protein dimer and a catalytic subunit termed DNA-PKcs are known to form the DNA-dependent protein kinase (DNA-PK), which binds to the ends at DNA double-stranded breaks to activate their repair. Her laboratory generated "knock-out" mice deficient in the individual components of the DNA-PK complex, and these mutant mice were then used to elucidate the roles of the different components in DNA repair, VDJ recombination, tumor suppression, and radiosensitivity modification.

The modification of the radiation response by Ku70/Ku80 also led her to explore the possibility of radiosensitization of cells through a reduction of intracellular Ku70 protein concentration, or through overexpression of a fragment of Ku70 that competes with intact Ku70 subunit in DNA-PK for DNA end binding. Subsequently, she devised animal models containing quadruple reporter genes for studying hypoxia-induced gene expression, for non-invasive imaging of tumor hypoxia and for hypoxia-targeted gene radiotherapy. Using the animal models her laboratory had generated, *in vivo* hypoxia gene signatures were obtained for both acutely and chronically hypoxic cells, and these signatures were shown to be of prognostic value in the analysis of clinical trial results.

During her career, Dr. Li has authored about 170 peer-reviewed papers and 30 chapters in books and conference proceedings. She has received continuous NIH/NCI funding as the principal investigator of numerous grants since 1981, some of which have continued for a period of two decades. She was accorded many honors over the years, including election to the presidency of the North America Hyperthermia Group, the Leonard Tolmach Visiting Professorship (Washington University), the Eugene Robinson Award Lectureship, and the Kallman Lecturer at Stanford University in 2013.

Michael Lovelock

Michael Lovelock received his Ph.D. in high-energy physics from the State University of New York at Stony Brook in 1990. He joined the Computer Services to work for Dr. Radhe Mohan on the software development of a specialized water tank dosimetry system that could be synchronized with the scanning beam of a new MM50 racetrack microtron. After using the system to commission the MM50, he worked for several years on the development of the 3D treatment planning system, TopModule. In 1995, he joined the Dosimetry Group to work with Dr. Tom LoSasso. At this time, MSKCC was opening centers on Long Island and New Jersey, so the initial clinical focus was the commissioning of all the treatment machines at the new centers for the TopModule planning system.

An opportunity arose to participate in the development of an image-guided treatment facility. Working with colleagues Drs. Kamil Yenice and Wendell Lutz, a facility equipped with an in-room CT scanner was developed and pressed into clinical service. The ability to accurately deliver radiation to sites outside of the brain made possible a new radiotherapeutic strategy for the treatment of spinal metastases. A long and ongoing collaboration with radiation oncologist Dr. Josh Yamada led to the development of the image-guided spine radiotherapy program at MSKCC. A radical approach advocated/championed by Dr. Zvi Fuks, the delivery of 24 Gy in a single fraction, was implemented and found to be highly successful, with local control rates of more than 90% reported. Failure analysis revealed the importance of the minimum target dose to local control. Michael worked on the development of patient immobilization systems and the quantitative assessment of dose delivery accuracy necessary for such treatments.

His current research interest is the clinical trials of electromagnetic guidance systems to facilitate the use of stereotactic body radiotherapy, in particular, the development of workflows that guide therapists in the management of the flood of information from real-time target position-tracking technologies. An associated interest is the clinical development of accurate and automatic quality assurance tests of the dose delivery and imaging systems. He has authored or co-authored over 70 papers and 12 book chapters.

Gig Mageras

Gig Mageras received his Ph.D. in physics from Columbia University in 1983. After a postdoctoral study in experimental particle physics at the Max Planck Institute for Physics and Astrophysics in Munich, Germany, he joined the Department of Medical Physics at Memorial Sloan Kettering

(MSK) in 1988. At that time, he was recruited as a research physicist funded by an industrial grant to investigate the application of a 50-MV racetrack microtron for computer-controlled radiotherapy.

He is currently an attending physicist in the Department of Medical Physics at MSK. He is also chief of the Medical Physics Computer Service, where he directs a team in the development and support of software systems for radiation treatment planning and verification. These systems are used in clinical and research activities by the Departments of Medical Physics and Radiation Oncology.

Gig has been active in numerous clinical initiatives in radiotherapy physics during his 30-year tenure at MSK. These include the development of a treatment management system (aka record-and-verify system) for computer-controlled treatments on the racetrack microtron; development of a radiotherapy PACS (picture archiving and communication system) to support electronic ("filmless") imaging for treatment verification; and introduction of deep-inspiration-breath-hold and respiratory-gated treatments in lung and abdominal disease sites.

In research, Gig has been a project leader and principal investigator on various NIH-funded and industrial research grants to study 3D conformal, intensity-modulated, and image-guided radiotherapy. His research team was one of two groups that independently pioneered a method of respiratory-correlated "4D-CT" for helical CT scanners, which was commercially adopted in Philips CT scanners and has gained widespread use for radiation treatment simulation. He also led the development of a respiratory-gated cone-beam CT technique for image-guided radiotherapy, which has been implemented by Varian in its TrueBeam linear accelerators.

Dr. Mageras has authored and co-authored over 100 peer-reviewed articles and 26 book chapters in medical physics and has mentored 18 postdoctoral fellows. He is currently a principal investigator on an NIH-funded R21 grant to investigate artificial intelligence methods to localize organs-at-risk in cone-beam CT scans and calculate dose received from ablative radiotherapy of pancreatic cancer.

Gig has been active in the American Association of Physicists in Medicine. He co-chaired an AAPM task group on the management of respiratory motion in radiation therapy. He served as therapy track coordinator and co-director of the scientific program for two AAPM annual meetings. For six years he was chairman of the Greenfield/Daniels awards committee for best papers in the *Medical Physics* Journal. Since 2006 he continues to serve as an associate editor for *Medical Physics*. In 2006 he was awarded Fellow of the AAPM.

Radhe Mohan

Radhe Mohan received his B.S. and M.S. degrees in physics from Punjab University, Chandigarh, India in 1962 and 1963, and a Ph.D. in Theoretical Nuclear Physics from Duke University, Durham, North Carolina in 1969. He completed his post-doctoral training at Rutgers University, New Brunswick, New Jersey. Radhe started his medical physics career in 1971 as Assistant Physicist and Instructor at Memorial Sloan-Kettering Cancer Center in New York. He served as Chief of the Medical Physics Computational Service from 1980 through 1994 and rose to the rank of Attending Physicist and Professor with Tenure and Associate Chairman. During his tenure at MSKCC, Radhe made pioneering contributions in a wide range of areas in the field of radiation oncology physics, including the design, development, and applications of the first computer-controlled automated dosimetry system; the first radiation treatment monitoring and delivery system; mathematical optimization methods; accurate methods of dose computations based on non-local energy deposition and Monte Carlo techniques; a brachytherapy dosimetry and treatment planning system; accurate computations of dose distributions for internally applied radionuclides; image processing; and 3DCRT and IMRT planning and delivery systems.

Radhe left MSKCC in 1996 to become the Director of Radiation Physics at the Medical College of Virginia, where he established one of the nation's first research and clinical IMRT programs. He led the team that investigated and developed 4D-CT imaging and the flattening filter-free (FFF) treatment mode of linear accelerators (now implemented in most commercial linear accelerators).

Radhe joined the University of Texas MD Anderson Cancer Center, Houston, TX as Chairman of the Department of Radiation Physics in January 2002, stepping down in October 2010 to focus on proton therapy research and clinical service. Currently, he is a tenured Professor and the holder of the Larry and Pat McNeil Chair in the Department of Radiation Physics at MDACC. During his time as Chair, he continued his IMRT, 4D-CT, FFF, and Monte Carlo research and development activities and established a strong department-wide research program in these topics, as well as in adaptive image-guided radiotherapy and dose-response modeling. Since 2005, his activities have been concentrated on various physical, clinical, biological, and, more recently, immunomodulatory aspects of proton therapy.

Radhe has been the recipient of many NCI and industry grants throughout his career. Most recently, he was the Co-Principal Investigator of a completed five-year NCI P01 Program Project grant entitled "Optimizing Proton Therapy." This Program Project was renewed in 2014 for another five years and has the new title of "Improving the Clinical Effectiveness and Understanding of the Biophysical Basis of Proton Therapy." The knowledge resulting from this research is being translated into the development and evaluation of novel proton therapy treatment planning and delivery methods for clinical trials and practice to enhance the robustness and effectiveness of proton therapy.

Radhe has over 450 publications, including over 375 in peer-reviewed journals. He was the Senior Physics Editor of *The International Journal of Radiation Oncology, Biology, and Physics* from 2002 through 2011. He is an active member and a Fellow of both the American Association of Physicists in Medicine (AAPM) and the American Society of Radiation Oncology (ASTRO). He has and continues to serve on various councils and committees of both societies. He also serves on the external advisory boards of several organizations, including the CERN Medical Applications International Strategy Committee, the International Conference on Translational Research in Radio-Oncology, Germany's National Center for Tumor Diseases, etc.

Among the many awards Radhe has received during his career are the 2003 AAPM Edith H. Quimby Award for Lifetime Achievement in Medical Physics; the 2004 Allan M. Cormack Gold Medal of the Association of Medical Physicists of India; the 2010 Failla Memorial Award of the Radiological and Medical Physics Society of New York, the 2013 ASTRO Gold Medal, and the 2018 AAPM Coolidge Gold Medal.

Keith Pentlow

Lawrence N. Rothenberg

Lawrence Rothenberg was awarded his Ph.D. degree in Nuclear Physics from the University of Wisconsin, Madison in 1970, and his thesis was approved in 1969. He completed a brief postdoctoral fellowship jointly in nuclear physics under Heinz Barschall and medical physics under John Cameron at Wisconsin. He then joined the Department of Medical Physics at Memorial Sloan Kettering Cancer Center (MSKCC) in New York as an American Cancer Society postdoctoral fellow in February 1970. After completing clinical rotations in many areas of medical physics, he joined the Diagnostic X-Ray Physics section of the Department when a staff position became available in November 1970. In June 1971, he became Section Chief when Stephen Balter left for a position in Massachusetts.

Larry concentrated his research activities on image quality and dosimetry, with particular emphasis on mammographic imaging and computed tomographic imaging. He was invited to present lectures on these topics at several AAPM Summer Schools, RSNA Refresher Courses, and at various local and national and international radiology and medical physics conferences. He was very active in the National Council on Radiation Protection and Measurements (NCRP), chairing Scientific Committee 72 which produced NCRP Reports No. 85 and 149, and also serving on the NCRP Board of Directors. Larry was similarly active in AAPM by serving as President and Chairman of the Board (1998–1999), in ACMP by serving as Chairman of the Board of Chancellors (1988), and in ACR by serving as the first physicist on the Council Steering Committee. He was a member of ICRU committees that produced reports on phantoms for radiology (ICRU Report 48) and for ultrasound imaging (ICRU Report 61). He was also a member of the ACR Physics Committee, which developed the ACR Mammography Accreditation Program, and he serves as a physics reviewer for the ACR Computed Tomography Accreditation Program.

For over 35 years, Larry served as the head of Diagnostic X-ray Physics Section of the Department and attained the position of Attending Physicist. He worked closely with Gian Ragazzoni and Lowell Anderson in teaching numerous topics in Diagnostic Physics to residents from Cornell Medical College (where he held an academic appointment), MSK medical physics fellows, and radiology residents and fellows from MSKCC and several neighboring institutions. After Lowell and Gian retired, Larry took over direction of both the physics lecture course for first-year residents and the physics laboratory course for third-year residents. In 2007, he received the Cornell University Medical College Radiology Department's Award for Excellence in Radiological Science Education, subsequently named in his honor. He has trained numerous MSK postdoctoral physics fellows in the techniques of diagnostic x-ray clinical evaluations.

Larry is currently Member Emeritus at MSKCC. He continues to serve the Department as Program Director for the CAMPEP Accredited Residency Training Program in Medical Imaging Physics, for which he received funds from AAPM to establish the program. He is also Associate Director of the CAMPEP Accredited Residency Training Program in Therapeutic Medical Physics, and Postdoctoral Physics Training Program Director.

Ellen Davis Yorke

Ellen Yorke received her Ph.D. degree in physics at the University of Maryland College Park in 1967. After a one-year postdoctoral fellowship in biophysics, she joined the physics faculty of a new and mostly undergraduate campus, the University of Maryland Baltimore County (UMBC), where she taught all areas of undergraduate physics, including the 'service' courses for premedical and allied health students. She first encountered the intriguing field of medical physics in the early 1980s through a course description in the *American Journal of Physics*, a publication aimed at college and high school physics teachers. She contacted Clifton Ling, who was then the Chief Physicist in the Division of Radiation Oncology at George Washington University (GWU) Medical Center. They applied for and received a Public Health Service award, through which Ellen was trained by Clif from 1983–1984. In 1985, Ellen took the leap and became a full-time therapy medical physicist in the GWU. With its two linacs, five attending radiation oncologists, three residents, two physicists, and a dosimetrist, the GWU had an active and innovative program in external beam radiation therapy and LDR brachytherapy.

Although Clif had moved on (first to UCSF and then to MSK), Ellen continued to collaborate with him on several projects that had begun during her training year, and she spent her sabbatical leave at MSK. During that time, Clif and Jerry Kutcher introduced her, together with Andy Jackson (then a post-doctoral fellow at MSK) to outcomes modeling, an area in which she has had great interest ever since. From 1997 through mid-1998, she was Chief of Clinical Physics in the Department of Radiation Oncology at the Hospital of the University of Pennsylvania and then came to the external beam Treatment Planning group at MSKCC, where she works today. In addition to studying radiation therapy tumor control and complications outcomes, she has been active in the clinical implementation of respiratory motion control in radiation therapy and in the 'patient safety' movement.

She has been active in AAPM, having served on several task groups— including being co-chair of Task Group 100 on applications of risk analysis to radiation therapy quality management—and was Chair of the AAPM Therapy Physics Committee from 2006–2011. She, together with Rock Mackie, initiated the AAPM/ASTRO-backed QUANTEC project, an interdisciplinary effort that abstracted dosimetric information from peer-reviewed publications of the "3DCRT" era to limit normal tissue complications from radiation therapy.

Marco Zaider

Marco Zaider received his Ph.D. from Tel Aviv University in 1976 for work in the physics of pions. For the next three years he was a post-doctoral fellow and staff member at Los Alamos National Laboratory. There he worked on the clinical application of negative pions to the treatment of cancer. In 1979, Prof. Rossi invited Dr. Zaider to join the Columbia University's College of Physicians and Surgeons, where he rose through the academic ranks to become Professor of Clinical Radiation Oncology, of Public Health (School of Public Health), and of Applied Physics (School of

Engineering). In 1988, Dr. Zaider became director of the graduate program in medical physics at Columbia University.

In 1998, Dr. Zaider was appointed Attending and Head of Brachytherapy Physics at Memorial Sloan Kettering Cancer Center and Professor of Physics in Radiology at Cornell University medical school, positions which he held until 2016, when he was appointed Emeritus Professor at MSKCC. Dr. Zaider served in leadership positions in many national and international scientific commissions (NCRP, ICRP, AAPM).

He is currently councilor in physics for the Radiation Research Society and a member of the main council of the NCRP. Dr. Zaider is board certified by the American Board of Medical Physics in radiation oncology. He has published two books and over 200 papers in peer-reviewed journals.

Others Involved

Howard I. Amols

Howard Amols grew up in New York City and received his B.A. in Physics from Cooper Union in 1970 and his Ph.D. in Nuclear Physics from Brown University in 1974. He then completed a two-year National Cancer Institute postdoctoral fellowship at Los Alamos National Laboratory, where he worked on the world's first negative pi-meson cancer treatment project. Upon completion of his fellowship he continued working at Los Alamos, first as a staff member, and then as an Assistant Professor of Radiology at the University of New Mexico Cancer center in Albuquerque, where he divided his time between pi-meson radiotherapy at Los Alamos and conventional x-ray and electron therapy at the university's cancer center. While there he was awarded an NCI Young Investigator grant to study microdosimetry of x-ray and electron radiation therapy beams. From 1977–1978, he was a guest scientist at Karlsruhe Nuclear Research Center in Germany, where he worked on pi-meson radiotherapy at the Swiss Institute for Nuclear Research in Zurich.

In 1981, he accepted a position as Assistant Professor of Radiation Medicine at Brown University and Rhode Island Hospital in Providence, RI where he continued his NCI-funded work on microdosimetry of radiation therapy beams. While at Brown, he was promoted to Associate Professor and Chief of Medical Physics.

In 1986, he became Chief of Radiation Therapy Physics at Columbia University in New York, rising to full Professor in 1991. While at Columbia he was instrumental in development of the tri-state area's first Stereotactic Radiotherapy Center. He also did pioneering work with Dr. Judah Weinberger on Intravascular Brachytherapy for the treatment of coronary artery restenosis, which is still used today as a salvage treatment modality for this disease. In addition, he was a co-founder of New York's first CAMPEP-approved medical physics training programs, many of whose graduates now work at Memorial Sloan Kettering as well as other prestigious cancer centers throughout the country.

In 1998, he was recruited to Memorial Sloan Kettering by Dr. Ling to succeed Dr. Gerry Kutcher as Chief of Clinical Medical Physics. His tenure as chief oversaw the growth of Memorial's network of satellite radiation therapy centers, as well as the continued development of intensity-modulated and image-guided radiation therapy, including the installation of North America's first Varian Truebeam linear accelerator. He was a senior investigator and project leader of the department's National Cancer Institute Program Project Grant as well as the department's industrial program grant from Varian Corporation. When Dr. Ling retired as Department Chairman in 2009, Dr. Amols succeeded him as Principle Investigator on both of these grants. He also supervised the training of numerous post-doctoral fellows and residents. For over 10 years he was also the principle lecturer for the Radiation Oncology residents physics course and was awarded Teacher of Year award by the residents in 2004. During the final years of his tenure as Service Chief, he directed early physics design studies for what eventually became the New York Proton Center. Dr. Amols stepped down as Service Chief in 2013 and retired in 2014.

His other service to the medical physics community includes service on the Executive Committee of the American Association of Physicists in Medicine (AAPM) as President Elect, President, and Chairman of the Board (2004–2006), as well as AAPM Liaison to the International Atomic Energy Agency. He has also served on the examination committee of the American Board of Medical Physics and on the RAMPS Raphex (Radiation Physics Examination for Radiation Oncology Residents) committee for nearly 30 years, being senior editor of the exam for many years.

For his services to the profession of medical physics, he was made an AAPM fellow in 1999, and he was awarded the AAPM Quimby Lifetime Achievement Award in 2014. In retirement, he continues to be active in the Columbia University CAMPEP training program as a Senior Lecturer, and he serves on several AAPM committees, plus the RAMPS Raphex committee.

Kristen Zakian

Kristen Zakian received her Ph.D. in bioengineering at the University of Pennsylvania in 1992. Her thesis advisor was Peter Joseph, Ph.D., an alumnus of the Memorial Sloan-Kettering Medical Physics Fellowship Program. That year, Kristen came to MSKCC as a postdoctoral fellow in Medical Physics. She joined the Imaging and Spectroscopic Physics Laboratory directed by Jason Koutcher, and she also rotated through the diagnostic radiology, nuclear medicine, and radiation safety physics services. After completing her fellowship, Kristen was appointed Instructor, and she divided her research time between performing clinical magnetic resonance spectroscopy research with Dr. Koutcher and developing novel radiofrequency resonators in collaboration with Dr. Douglas Ballon, who was an attending Physicist at MSKCC for many years and who now directs the Citigroup Biomedical Imaging Center at Weill Cornell Medical College. Kristen and Doug Ballon share a patent for two-dimensional ladder network resonators. Subsequently, Kristen was appointed Assistant, and then Associate Attending Physicist, in the Department.

Following her initial collaboration with Jason Koutcher, Kristen eventually focused her research on clinical and preclinical investigations of MR spectroscopy in cancer as a marker of tumor prognosis and response to therapy. She has headed or collaborated in multiple grant-funded studies employing phosphorus-31 or proton spectroscopy for the prediction of treatment response in prostate cancer, sarcoma, colorectal cancer, and lymphoma. In addition, she developed an interest in liver metabolism during regeneration and chemotherapy, and she obtained grant funding to study the effect of chemotherapy on liver metabolism in patients with colorectal cancer. In connection with her ongoing studies, Kristen has acted as a mentor to graduate students and postdoctoral fellows who have been involved in the research.

Kristen's other responsibilities include supporting the MR section of the Small Animal Imaging Group, including sharing her expertise on radiofrequency coil construction, data acquisition, and quantification of metabolites from *in vivo* spectroscopy data. With the acquisition of the first MR Simulator in the Department of Radiation Oncology, Kristen has become one of the supporting physicists and is planning to pursue research linking MR techniques to radiation-therapy-related questions. Kristen teaches MR physics in various lecture series for residents and fellows in Molecular Imaging, Radiology, and Radiation Oncology. She also provides analysis software and support for radiologists using multimodality MRI for the investigation of treatment effect.